Building a
Network

Titles of Related Interest

Bates, Wireless Networked Communications

Black, Network Management Standards, 2/e

Emery, How to Be a Successful Systems Manager

Enck/Beckman, LAN to WAN Interconnection

Feit, SNMP

Jayachandra, Reengineering the Networked Enterprise

Johnston, OS/2 Connectivity and Networking

Minoli, 1st, 2nd, and Next Generation LANs

Minoli, Analyzing Outsourcing

Naugle, Local Area Networking, 2/e

Naugle, Network Protocol Handbook

Peterson, TCP/IP Networking

Sheldon, LAN Times Encyclopedia of Networking

Szuprowicz, Multimedia Networking

Terplan, Effective Management of Local Area Networks

Udupa, Network Management Systems Essentials

Woodcock, Networking the Macintosh

Building a Network

How to Specify, Design, Procure, and Install a Corporate LAN

Peter D. Rhodes
PR&A
Derry, New Hampshire

McGraw-Hill

New York   San Francisco   Washington, D.C.   Auckland   Bogotá
Caracas   Lisbon   London   Madrid   Mexico City   Milan
Montreal   New Delhi   San Juan   Singapore
Sydney   Tokyo   Toronto

Library of Congress Cataloging-in-Publication Data

Rhodes, Peter D.
    Building a network : how to specify, design, procure, and install
a corporate LAN / Peter D. Rhodes.
        p.   cm.
    Includes index.
    ISBN 0-07-052134-4 (acid-free paper)
    1. Local area networks (Computer networks) 2. Management
information systems. I. Title.
TK5105.7.R48   1996
004.6'8'0684—dc20   95-21559
                                      CIP

McGraw-Hill

A Division of The McGraw-Hill Companies

2 3 4 5 6 7 8 9 0 BKP/BKP 9 0 0 9 8 7 6

ISBN 0-07-052134-4

*The sponsoring editor for this book was Marjorie Spencer, the editing supervisor
was Paul R. Sobel, and the production supervisor was Donald F. Schmidt. It was
set in Palatino by Ron Painter of McGraw-Hill's Professional Book Group compo-
sition unit.*

Printed and bound by Quebecor/Book Press.

McGraw-Hill books are available at special quantity discounts to use as
premiums and sales promotions, or for use in corporate training programs. For more information, please write to the Director of Special
Sales, McGraw-Hill, 11 West 19th Street, New York, NY 10011. Or contact your local bookstore.

 This book is printed on recycled, acid-free paper containing
a minimum of 50% recycled de-inked fiber.

This book is dedicated to my parents—a little more vigorish

Contents

Preface   xiii

1. Introduction   1

Target Audience   1
  Who Are They?   2
  Skills Required   3
Tools Needed   5
  Hardware   6
  Software   6
Needs to Be Filled   7
  New Network   8
  Addition to Current Network   8
  Scrap and Rework   8
Four-Step Process   9
Preparation   9
  Who Is Involved   10
  Regulatory Requirements   11
  Consultants and Others   12
Magnitude of Efforts   12
Summary   14

2. Specify   15

Result of Specification Is a Design   16

Functions, Not Brand Names   17
    Layered Approach   17
    Path Approach   18
Sources of Data   20
    Bottom Up   20
    Survey and Questionnaire   23
Analysis of Data   24
Review   25
    Rough Outline of Requirements   26
Identify Critical Mission Requirements   28
    Single Point of Failure   28
    Review   32
What Is the Business?   32
    How Will The Network Support the Organization?   33
    How Will Daily Operations Change?   34
Growth of The Organization   35
Network Management System   36
    What Reports Are Needed   37
    Network Management Systems Expertise   38
    Responsiveness   38
Review Phase   39
    Types of Questions for Subscribers   40
    Who to Review With   40
    Practice Defensive Management   40
Summary   41

3. Design   **43**

Analysis by Throughput   44
    By Business Function   45
    Analysis by Location   47
Physical Distances Involved   49
    LAN Extensions   49
    WAN Extensions   51
Choice of Transmission Media   51
    Copper Media   51
    Fiber-Optic Media   54
    Radio as a Transmission Medium   56
    Free Space Light   58
    Review   59
Connectivity   59
    Proprietary versus Nonproprietary   60
    Open Systems Interconnect   61
    IEEE Protocols   63
    Review   66
    Network Operating System   66
Network Extension Hardware   70

Considerations 70
Decision Matrix 71
Inter-networking 74
Failure Mode 75
Review 75
Test Plan 76
Must Have 76
Nice to Have 77
Physical Layer Testing 77
Network Layer Testing 79
Transport Layer Testing 80
Application Layer Testing 80
Testing Considerations 81
Summary 81

4. Request for Quotation 83

Where Does the RFQ Come From? 83
The RFQ Is a Legal Document 84
Legal Practitioners 84
The RFQ Provides Guidance 85
What Is to Be Delivered 86
When Is It to Be Delivered 87
Quality of Materials to Be Used 87
Quantity of Material to Be Used 88
Unit Pricing 89
Contents of the RFQ 89
Administrative Information 89
Broad Description 91
Hardware and Software Specifications 91
Installation 96
Test and Acceptance 98
Qualifications of Bidder's Staff 99
Financial Qualification of Bidders 99
Acceptance Form 100
Appendixes and Annexes 100
Change Methodology 100
Summary 101

5. Analysis of Responses 103

Evaluations Must Be Based on Objective Data 103
Intermediate Results 105
Unit Price 108
Cost of Labor 109
Selection of Semifinalists 109
Project Management Plan 111

Best and Final Offer   113
Summary   115

6. Contractual Documents and Legalities   117

Legal Advice Necessary   118
A Contract Provides Protection to All Parties   119
Financial Considerations Are Contractual   120
  Escrow   120
  Bonds   121
Remedies for Performance Failure   121
Types of Law   123
  Uniform Commercial Code   123
  Federal Regulations   123
  National, State, and Local Codes   123
Ownership of Material   124
Insurance   124
Property Insurance   125
  Workman's Compensation   125
  Professional Liability   126
  Insurance for Gaining Organization   126
Subcontracting   127
Summary   127

7. Installation   129

Preinstallation Activities   130
Inspection of New Material   135
Staging of New Material   138
Contractor's Employees   140
Working Hours   144
Safety Hazards   145
Notification of Completion   146
Preventive Activities   147
Who Tests?   148
What Is Tested?   149
Test Results   150

8. Evaluation of the Effort   151

Excuses and Other Weasel Words   152
  "Yes, But..."   152
  Brain Cramp   154
  We Tried the Best We Could   154
  They Wuz Outa Stock   155
What Documentation Is Required?   157

      Forms and Records   157
      Payment and Release   158
  Summary   158

9. Project Management   **161**

The Project Management Triangle   161
Critical Path Method   164
  Beginnings   164
  Work Breakdown Structure   164
  Project Estimate   166
  Next Step   167
  Predecessor/Successor   167
  Cross Unit Relationships   168
  Final Step   168
  Milestones   168
  Review   169
Issues   169
  Levels   169
  Issue Work-Off Procedures   169
  Issue Work-Off Completion   171
  Issue Review Team   172
  Review   177
Reports and Project Management   177
  Costs   177
  Delay   178
  Legal Actions   178
  Lost Time Injuries   179
  Material Deliveries   179
  Milestones   179
  Periodic   179
  Phase Completion   180
  Property Damage   180
  Shrinkage   181
  Subcontractor Activities   181
  Test   181
  Turnover   182
Summary   182

Appendix A. Organization Description   **183**

**Appendix B. Bent Metal Corporation Chassis
    Builders to the Electronic Industry**   **193**

Appendix C. Contract for Network Installation   **209**

  Index   219

Preface

This is where we start, with a few, very brief paragraphs about beginnings and their impact on endings.

First, this is a companion book to *LAN Management, A Guide To Daily Operations*. In that book we expect you already have a network. Here we will tell you how to get one if you either do not have one, or are unhappy with your present one.

Like our companion book, you will not need a degree in computer science, physics, or mathematics to understand what is going on. Simple, grade school math including addition, subtraction, division, and multiplication are all that is necessary. A reasonably flexible computer which can run common spreadsheets, word processing, and graphics or computer aided design (CAD) packages is a definite plus. In all reality, the tasks could be done with pencil and paper, but if that approach is selected, why not step even further back to quill pen and parchment?

The final network will depend on the decisions made now. Period—end of statement! Your dream will be realized in copper, glass, steel, and silicon. A mistake now will be something you have to live with for as long as you work with the network. Of course, if the mistake is too severe, your discomfort will not last very long.

In the realization of your efforts you will alienate people. Know this and expect the slings and arrows of outraged subscribers, department heads, division heads, and probably the janitorial staff as well. You will irritate everyone; prepare for it and then get used to it. It will happen. There are ways to ameliorate this irritation. Although this book is not the proper vehicle to discuss team building, the best approach to minimize the irritation is to maximize other subscribers inputs into the specification and design phases of the activity.

There are four steps to creating a successful network—specify, design, build, and test; these are defined and discussed later. The key point to realize is that these are all parts of the same whole, and are indivisible.

Finally, be careful of others. Vendors are glad to provide design skills; as long as their products are the ones being bought. Cable installers will design the copper and fiber paths, but in such a way as to maximize their profit, not the final flexibility. Consultants who do not sell hardware will help, but they make their money selling time and skills. Get a firm, fixed price for their efforts. Such actions will get this type of consultant out of your hair as soon as possible. Consultants who sell hardware or software have an axe to grind. Be careful that their plans do not rely solely on their hardware. Otherwise, that person is a representative from some company who does not know they have a "secret admirer." A final warning is needed here—free advice is worth what you pay for it. If you ask a reseller if their product will fill your need, do you think you will hear anything but how good it is?

Attached is the organization chart for a fictitious firm known as the Bent Metal Corporation. This firm will be used to simulate an organization which is moving from a mainframe/terminal paradigm to a distributed computing/LAN paradigm.

Still with me? Good! Continue reading and you will be on your way to an understanding of what you are starting. Lucky you!

Acknowledgment

Technology, by itself, is little more than a laboratory curiosity. The application of technology to human endeavor is a high calling, one to be proud of. This book acknowledges the mostly invisible efforts of those men and women who daily apply telecommunications technology to the entire spectrum of human endeavor. Thank you for a job well done.

Peter D. Rhodes

1
Introduction

Begin at the beginning . . . and go on till
you come to the end: then stop.
LEWIS CARROLL
Alice's Adventures in Wonderland (1865)

This chapter provides the groundwork for the remainder of the book. It addresses

- The types of people and organizations who will gain the most from the book
- The tools necessary for a proper network procurement
- An analysis of the procedures to be used to determine the functional characteristics and design of the network
- The four-step procurement process
- Certain preparatory processes
- The impact of nonfinancial leverage in the network procurement process

Target Audience

This work is aimed at the individual who, in any organization, is charged with the responsibility of modifying an existing network, or less likely, installing a new network. This does not require that the individual be the project manager or even technically qualified to perform such a function. In an organization of fewer than 100 employees, the networking "guru" is the person who knows how to install a local area network (LAN) card and connect it to the network. Discussion of standards, topologies, and topographies probably will draw a glassy stare and a shake of the head followed by "Uh, well . . . let me get back to you on that."

In an organization of more than 1500 employees, the reverse may be true, but the person charged with the installation task is a manager who probably knows less than the most recently hired analyst concerning what is and is not a good network. At least the analyst has some built-in bias toward some proprietary or nonproprietary standard, and thereby has an understanding of it. This nontechnical manager will delegate the task to the engineers or subordinate technical managers who know what they are doing, and can, through their advanced technical knowledge, "snow" their nontechnical manager into accepting technically absurd positions. Think it cannot be done? Ask a technician to explain his or her specialty to you when you do not understand it. Do not forget: The nontechnical manager is the decision maker, not the engineer who designs and installs the network.

Who Are They?

Neither of these two extreme examples is the prime audience for this book. Rather, this book is aimed at the person who has some knowledge of the network currently running in the organization and has been charged with changing it to reflect the new competitive reality the firm faces. Recently this has been called downsizing, rightsizing, or distributed computing. Whatever term is used, the organization is trying to hide the fact that what is there now is not sufficient for today's challenges or circumstances.

This book is written for someone with some technical knowledge, perhaps an engineer (hardware) or computer scientist (software). He or she has installed, maintained, and managed the network for several years and has a good feel for it; not only at an intellectual level, but at a gut level as well. More importantly, this person understands the organization and how the network in use now supports it. Further, this person understands how the network planned for the future will affect the way the organization does business today and tomorrow. For the purpose of the rest of this discussion we will call this person the Network Manager.

Skills Required

The Network Manager needs several different types of skills, not all of them technical. However, let us examine the technical skills first.

The Network Manager must be able to evaluate both proprietary and open technologies and their applicability to the organization. One of the questions that may have to be answered is something like: Is a ring topology better than a bus topology, and if so, why? If not, why not? This is the first step on a long and difficult road. Once the topology is selected, it must be sold to upper management, middle management, and subscribers. How difficult will it be to convince someone who has used only Digital Equipment Corporation products that some NoName Startup, Inc.'s products can pass bits just as well as DEC's can? In counterpoint, this decision may not even be a final one. The Network Manager may gain certain information through analysis that indicates a need for a greater throughput, in bits per second, than the selected topology provides. We now have identified two necessary skills, selecting topology, which is the form of the network, and persuasion.

The Network Manager must also be able to select those people within the organization who will be capable of supporting the new network, either through existing skills or by being retrained in the new topology. If the Network Manager recognizes none of the current employees are competent in the new topology, then they must either be retrained from the ground up, transferred elsewhere within the organization, or terminated. The Network Manager must therefore be a people manager as well as a technology manager.

Another consideration is that the Network Manager will be constrained by budgetary realities. The new network does not appear by magic, it will be constructed by people employed either inside or outside the organization. These people will have to be paid for their time and materials, at a minimum. The money for this effort must be budgeted, probably well ahead of the actual construction date. This means that the Network Manager must be able to determine how much the new network will cost before it can be designed. Huh? Say that again please! If the Network Manager is designing a network in order to figure out how much it will cost, how can he or she determine how much to budget *before* it is designed, since it cannot be designed until the Network Manager knows how much it will cost? Circular reasoning strikes again. This problem appears worse than it actually is. We will demonstrate that the Network Manager knows fairly well how much the new network will cost before it is more than a dream on the distant horizon. As such, the Network Manager also must have the skill to be able to complete budgets.

Negotiation skills are necessary for the Network Manager to ensure the new network is a useful product. Vendor negotiation will not be mentioned, it is assumed. The Network Manager must also negotiate with users for bandwidth requested versus the amount of money available to purchase the technology. Negotiation is needed with those charged with upkeep on the buildings, or with the landlord if the physical plant is leased. Negotiations may be required with representatives of local exchange carriers (LXC) and interexchange carriers (IXC) for tariffs, maintenance costs, and installation dates. The Network Manager may also have to negotiate with consultants for their services. (We identify consultants as those who provide skilled services only; all others are vendors, selling theirs or someone else's hardware or software.) Although some may not see it as negotiation, the Network Manager must work with junior members of the networking staff. These people must be persuaded that the upcoming changes are for their benefit, even if it means they will be out of a job in the future.

The Network Manager must have marketing skills. The marketing effort can consist of providing questionnaires, taking surveys or initiating other formal information-gathering methods. The results of this information gathering will be an understanding of the subscriber's perception of their needs rather than the Network Manager's perception of the organization's needs. These results will only come from an analysis of the answers given by the subscriber community. The second aspect of necessary marketing skills includes the ability to sell a particular plan of action based on the analysis of the subscriber community. These persuasive skills were discussed above.

The Network Manager also needs to be able to manage the new network once it is installed. This requires skill in the selection of network management packages with an eye to future changes in protocols, topologies, and standards. So in addition to understanding the operational hardware and software involved with the new network, the Network Manager must understand new network management hardware and software. Keeping abreast of topologies and standards is an activity that can consume over eight hours in any given week, by merely reading the trade publications and published positions of standards committees.

Finally, the Network Manager must be knowledgeable of the existing codes, ordinances, and rules governing the installation of wire and cable within buildings having public or employee access, such as an office or manufacturing environment. Many of these regulations are in publicly available documents. Certain fellow employees in the facilities or custodial departments also may be a source of knowledge in this area.

In summary we have identified the person who will gain the most from this book as one who is charged with managing the day-to-day operation of the existing network, or planning for an original network in an organization that never had one in the first place. And we have said that the person must have skill and/or knowledge in the areas of

- Technology evaluation
- People management
- Budgeting
- Negotiating with vendors and in-house persons
- Marketing
- Selection of network management hardware and software
- Local codes and ordinances

Once the Network Manager has polished these skills, this same person must select the tools used to implement these skills to start the specification stage.

Tools Needed

The Network Manager needs several generic types of tools, hardware, and software running on a personal computer of some sort. But the most important tool is the plan of action. The plan is produced without the use of anything other than the Network Manager's own abilities and experience.

The plan for network procurement and installation does not include types of hardware, software, or any other kind of -ware. The plan is nothing more than a series of steps and the sequence in which they will be performed. This four-step process is diagrammed in Figure 1-1.

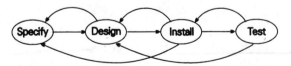

Figure 1-1. The four step process.

Note the feedback loops between Specify–Design and Install–Test as shown in the figure. These are iterative steps which feed on each other, resulting in a product that has gone through many, many levels of examination before the first dollar is spent. This four-step process is the heart and soul of the effort and is discussed in several ways below.

A subset of the overall plan exists between the design and install phase. This is where the Network Manager actually decides when the work starts, how long the selected bidder has to perform the work, and other variables concerned only with the actual, physical installation. This is the portion of the effort that will be reduced to Gantt charts or PERT diagrams, depending on the Network Manager's preference and the size of the effort. This may be safely ignored if the complexity of the installation can be handled on the back of an envelope.

Next, there is a section addressing this four-step process. Following Chapter 1, each of the four steps is assigned a chapter of its own, where each step is analyzed in detail. But first, the necessary tools should be examined in detail.

Hardware

The Network Manager needs certain tools, the first of which is hardware. A computer of some type is needed to record, store, and manipulate data derived from the analysis of the current situation. The larger the amount of data involved, the larger the memory required, and the larger the hard disk required. Rapid access to and manipulation of the data involved requires a fast single processor or parallel processors coupled with high clock rates. A printer capable of reproducing a single 8½-by-11 inch drawing with reasonable rapidity is mandatory. (A worst case speed is four pages per minute.) A printer built around laser or ion deposition technology is almost mandatory. A plotter, capable of up to E size prints is a nice option. A scanner capable of scanning drawings for storage on the computer is an advantage as well. Access to a facsimile machine may be mandatory during the finalization of the bid. If others have computer communications capabilities, a modem with a separate telephone line also may be useful.

Software

The required software falls into three general categories. The first is either a spreadsheet or a database program, whichever the Network Manager finds easier to use. When planning a network, much information will be received, stored, and manipulated. The framework provided by a database or spreadsheet makes assimilating this information easier than filing many pieces of paper. A word processing package, which is capable of generating multiple-page documents which themselves are subject to many revisions, is the key here. The same package will have the capability to include simple line drawings in the text. A graphics or CAD package for the creation of anything from simple line drawings to E size drawings (blueprints) will be most helpful. By providing the potential bidders with pictures, the Network Manager will rule out many differences in perception. If a modem is used, some type of telecommunications software is mandated.

To review, the Network Manager needs three things:

A plan

Hardware

   Computer

   Printer and/or plotter

   Modem

Software, including

   Spreadsheet or database program

   Graphics or drawing application

   Word processing

Once the hardware and software has been installed and is operating and the Network Manager is comfortable with their use, they will be used to complete the identification of the needs to be fulfilled. This process is as follows.

Needs to Be Filled

Data gathered from observation and the subscriber community, which is then evaluated in light of financial and physical realities, provide the answer to the question "What needs must be filled?" The Network Manager must decide whether

1. A new network will be installed (A network is nonexistent now),

2. The current network is to be expanded, or

3. The current network is to be scrapped and a new one installed.

New Network

Installation of a new network is something that happens only when a small organization recognizes the need for a network or a large organization moves into new quarters. Buildings that are wired according to the specifications of EIA/TIA standards 568 and 569 will eliminate many problems the Network Manager will face. Those buildings, however, are few and far between as of the copyright date of this book. The Network Manager may want to obtain a copy of these two standards as a guide to network design at the physical level. This information is available by contacting the

Electronic Industries Association

Engineering Department

2001 Pennsylvania Ave., NW.

Washington, DC. 20006

(202) 457-4966

One of the major points to consider when installing a new network is what numbering systems or means of identification are going to be used. In most networks, some type of electronic network address is required as well. These types of identifications must be recorded and kept up to date on a day-to-day basis.

Addition to Current Network

In counterpoint, if the current network is merely expanded or extended, then the number of decisions to be made by the Network Manager are very few indeed. Basically, the Network Manager need only answer questions concerning money and quantity of parts installed, followed by the anticipated dates of completion.

Scrap and Rework

The question of whether to scrap and rework the network that is currently in place is a difficult one in any condition. Since the network is in use now, one or more subscribers will be out of communications for some measurable period when their hardware is disconnected from the old network and reconnected to the new network. The duration between disconnection and reconnection is a judgment call made by the Network Manager. There is also a financial decision concerning the decision to scrap and rework. Has the existing network and all of its hardware and software completely depreciated? If not, the accountants will be only the first ones who will be calling the Network Manager asking how this "$light, but co$tly mi$take came to pa$$?"

Another question concerning the scrap and rework decision is also part of the problem associated with the need to add to the current network. If the current network is working so well, why change? On the other hand, why should the Network Manager continue fighting the same problems and suffering from the same restrictions every day?

Each of these problems will be addressed in more detail later, in Chapter 2, when the data-gathering process is complete and the analysis begins. But first, let us look at the four-step process used in network procurement.

Four-Step Process

The four steps involved in building a network as were mentioned above are to specify, design, install, and test the finished product.

The specification step is preliminary. The number of variables is large, there is a lot of fog in the air (as in "where t'fog are we?") and the result of specification is only a design. The design cuts through much of this fog, but in so doing shows many problems which will have to be dealt with in the very near future. The installation is the part that is most visible to the subscriber community, and it will show that community how much control the Network Manager has over the entire activity. If only the subscriber community could know the whole truth. The testing effort determines whether the organization gets its money's worth. For example, if the copper cable has a specified throughput of 100 megabits per second (Mbps), how does the Network Manager know this?

In partial preparation for the planning of the network, and before the specification process begins, the Network Manager must consider certain ugly realities of life beyond those of money and technology.

Preparation

There are three general areas of preparation to consider: what internal staff is involved, what additional labor is required, and what legalities are involved.

Who Is Involved

Does the expertise exist in-house for building a network? It is a rare organization that has the necessary skills in all the steps involved. The Network Manager, for example, may have experts in wire, but do they understand fiber construction techniques? Or, if a technician can install a bridge, can this same technician troubleshoot a router? If one of the bidders provides financial statements with the bid (a recommended procedure), who can analyze them for accuracy and correctness? Who in the Network Manager's staff is capable of preparing a PERT chart and following up with Gantt bars or creating a Critical Path chart for the installation?

Should the Network Manager bring in an outside consultant? The answer is yes, no, and maybe. Yes—if the need is present. This need will be obvious if there is a lack of in-house expertise. No—under some conditions, particularly if the effort is an extension of the current network. Maybe—is the answer when neither of the two previous examples are obviously present. A consultant is useful for performing "sanity checks" on plans which exist, but before the bidding process begins. A consultant also can be useful in the final, quality-control portion of the effort.

How does the Network Manager select a consultant? Good luck! Many of these types of persons have academic connections through local colleges and universities. A Network Manager can skim several years of trade publications to look for authors' names. If the same one crops up several times, this author may be a consultant. The same can be said for those who write books on the subject. Once several names have been identified, contact these people and see if they offer the types of services needed. If so, ask for references and then check the references given. One final word here: Hiring a consultant is not an admission of inability. Rather it is a statement that says the Network Manager knows the limits of his or her current knowledge and knows that an error overlooked will be dangerous. No one realistically complains about the cost of insurance, right? A consultant offers a type of insurance as an unbiased opinion on the Network Manager's plans and activities.

Are additional permanent staff members required? In all likelihood the answer will be yes, regardless of whether a new network is installed, the existing one is expanded, or the existing one is reworked.

A company's new network may require new skills of its personnel. Where these skills come from is the problem of the Network Manager. As stated above, the Network Manager needs personnel management skills. A network that has grown from its current base needs more people to support it. How many more is a question without a ready answer. Only the Network Manager knows how the current staff is assigned, how much slack time workers have, and how much overtime is necessary. A scrap and rework approach means that the Network Manager must either retrain the existing staff members or release them and procure a staff with a new skills set.

Although the Network Manager should work closely with the organization's personnel manager, there are several options the Network Manager should consider as a source of new or additional labor. The first source to consider is the possibility of recent technical school graduates. These people may be knowledgeable about the newest technology which the Network Manager is considering for installation. A second source is a temporary agency, which may provide a ready pool of technical skills. There are additional benefits to be gained from dealing with temporary agencies. The salary of the worker is high, but agency hiring fees are minimal to nonexistent. The Network Manager also gets a chance to "test drive" a prospective employee, evaluating him or her for the fit within the organization and the degree of technical knowledge the person brings to the position. The third source of potential employees is that of independent contractors. These hired guns do temporary work or job-shopping as if they were their own temporary employment agency. These people are particularly useful during the project management and installation phases of the effort.

Regulatory Requirements

Although people are part of the preparatory phases, the Network Manager must also consider the regulatory requirements affecting the new network. Internal corporate policy must be identified and evaluated for its impact on the installation effort. More importantly, the local laws, regulations, codes, and ordinances have a major impact on the installation process. The Network Manager has several options at this point in the preparatory phase.

The first option is to rely on the knowledge of the installers in the area of these regulations. This may or may not be a good idea. If the installers are local and have been in business for many years, they may know the regulations. If, on the other hand, the installers are from out of town or state or have been in business for just a few months, they may not be knowledgeable about the regulations and could lead the Network Manager astray. In most jurisdictions, the installers may be forced to extract improper cabling and reinstall it so that it meets the regulations. While this will not cost the Network Manager any money to do so, it will lose time and delay the start-up of the new or revamped network.

Consultants and Others

The second option to ensure the installation meets regulations is to hire those with the necessary knowledge. Who has this knowledge? There are several sources, including those who teach the regulations at local colleges and universities. Many of these people regularly provide courses for code enforcement officers, so they are very knowledgeable of any last minute changes. Retired code enforcement officers also may be a source of expertise, and there are lawyers who are skilled in handling the finicky details involved, but they can be expensive. Consultants may be appropriate, but we do not necessarily recommend them. Just how knowledgeable are these persons, and what else are they selling beyond their expertise?

In sum then, the Network Manager must make three preparatory decisions before the actual effort begins. These include:

1. Determining personnel requirements
2. Deciding whether consultants are required in the planning stages
3. Evaluating the regulatory requirements for the installation phase

There is one other strictly intellectual activity which is part the preparatory stages—an analysis of the effect of leverage on the installation procedure.

Magnitude of Efforts

It is a commonly demonstrated law of physics that leverage increases the available power of human muscle. This is demonstrated in the inclined plane, hydraulic jack, and pry bar. The same concept applies to the installation of a new network. A little effort at first provides a much greater result in the end. Some examples, please!

If a minor effort is made to check references, it is easy to weed out marginal bidders, thus providing a better installation in the end. Assume the Network Manager has 15 responses to the bid which was published. The number of responses must be reduced to five for a second round of bids. Why give a poor quality bidder a chance when it is possible to identify the lack of quality of service and products offered? The Network Manager eliminates the time-consuming activity of evaluating material specifications and dollar values quoted from those who are not qualified for the work. These marginal performers are then eliminated from the short list.

Early specification of components eliminates time-wasting studies of various vendors' offerings. By deciding Brand X wire is the best offering and Brand Y hubs are the most robust and Brand Z network access units are the most reliable, the Network Manager does not have to evaluate each bidder's presentation. The drawback to this approach, however, is that the Network Manager gets locked into a particular technology and cannot readily or easily react to recent changes in the marketplace.

Another example of a little effort producing greater results is that rank ordering of the final bidders makes picking a fallback vendor easy. If Vendor A drops out of the bidding at the last moment, it is not necessary to reevaluate all the bidders. Although not mentioned previously in sufficient detail, the concept of creating an objective set of data about vendors for their evaluation is a goal. The documents sent to the potential bidder should have point values assigned to all truly important areas. The more important the area, the higher the points granted. Make these decisions now, not when evaluating the replies from bidders. See Appendix B, The Request for Quotation.

It is quite difficult, if not impossible, to foresee all the problems and errors that will show up during the four phases of the procurement of the new network. The Network Manager may want to consider the following process of how to plan for those mistakes that can be foreseen.

1. Pick one person, preferably one with extensive knowledge of his or her department, from each department in the organization.

2. Get these people together on a routine basis and let them present scenarios of what could go wrong with the network. Stress that they are to use their imagination.

3. Let them provide multiple approaches to dealing with these potential, worst-case horrors.

4. Rank order the groups of solutions to these potential horrors by degree of applicability.

5. Be prepared to apply the solution(s) selected when the problem(s) arise.

Some problems based on our personal experience include a strike by transportation unions. The needed material cannot be delivered on the scheduled start or installation dates. A natural disaster causes a lack of electricity, leaving workers literally powerless and in the dark for days. Contractor(s) declare bankruptcy: What would happen if the vendor selected for your connectors suddenly went bankrupt and is not shipping any orders to the installer? Or, even worse, the installer goes "toes up"? Oh yes, and do not forget the ego problem. "What! You mean I wasn't consulted in the planning stages! Why not?!"

Summary

This introductory chapter is intended to give the reader a flavor for some activities that prepare the Network Manager for the creation of the new network. In this chapter we have

- Identified the people who will gain the most from this book.
- Touched upon the tools needed to begin the work.
- Noted the three options with a new network and the four-step process used to select and install it.
- Brought forward the preparatory questions for evaluation and analysis.
- Mentioned the difference in actions taken now versus actions taken later as leverage.

2
Specify

In dreams begins responsibility.   W. B. YEATS
  Responsibilities (1914), epigraph from an
                              old play

The result of the specification process is a design. But this is getting ahead of what must be done. In the specification process, many preliminary activities must be executed before the specification is complete. We will

- Discuss the relationship between function and brand names, and the trap which is not so evident there.
- Demonstrate the creation of the foundation for design specifications in the form of data gathering.
- Demonstrate that analysis of the data provides other, more informal, and sometimes unquantifiable information for the Network Manager's efforts.
- Discuss certain critical mission requirement activities, which must be evaluated above and beyond the subscriber's point of view.
- Point out the culture of the organization and how the new network will fit into that culture.

■ Mention that each network needs network management systems. Here we will discuss the impact of network management systems on the specifications for the new network, and how the network must be specified in terms of future growth.

Result of Specification Is a Design

There is a relationship between specification and design. That relationship was shown in the previous chapter as part of the four-step process. The design is a result of the specification, therefore errors in the specification will become evident during the design and installation phases when it is too late to be corrected at low or no cost. Subtle errors will not show up until subscribers begin to stress the network. Just because the network works under test conditions does not mean the network will work under real world conditions.

Although both Chapters 3 and 7 address testing of the new network, some mention of the concept of testing vis-a-vis real world operation is necessary. Each component of the network can be tested by itself and pass with flying colors, yet the network dies when it is turned on. The physical connectivity and logical connectivity may work, but the network will not pass traffic at anything much above 10 percent of the specified throughput. Or, the network can pass traffic quite well, but the software in use does not interoperate and constantly causes one or more servers to crash. This kind of testing, real world testing, cannot be done in a production environment. The more sophisticated the network, the more obscure the problems that come up. The more obscure the problems, the longer they will take to surface and the longer they will take to eliminate.

The only way in which these subtle problems can be eliminated is to create the entire network, as a network, in an empty building or unused space. Each component is tested by itself, then transmission media are connected. Next, all equipment is tested for interoperability. Finally, applications software is loaded and then tested again. This is very, very expensive and quite time consuming. The only gain achieved through this process is that the Network Manager will know if the network will work from the moment it is installed. This approach is recommended where the network is used for life support functions or law enforcement and military activities.

Functions, Not Brand Names

Getting back to the point of this discussion: The result of specification is design; one form of this design is a listing of functions. The following functions are representative only.

- Share spreadsheet data
- Pass documents from person to person in electronic format
- Share a common database
- Participate in E-mail
- Share expensive resources in the form of printers or high-speed Central Processing Units (CPUs)

These are general in nature and each organization has its own particular functions to add to the list. When the functions are identified, the Network Manager must select one of two broad approaches to the network specification. Additionally, the Network Manager must be aware of the trap of using brand names as a specification. By specifying Novell's Netware Network Operating System (NOS), for example, the Network Manager has also specified the protocols that will run on the network. This may not have been the desired result.

Layered Approach

The Network Manager has two tracks to choose from in the preliminary activities. The first track is somewhat like the Open Systems Interconnection (OSI) model, developing the network as a series of layers, each layer describing a subfunction of the main function shown above.

In Figure 2-1, the physical layer is the transmission media, used to provide connectivity between two pieces of hardware. The next layer

```
┌─────────────────┐
│ Data Manipulation │
│ And Storage      │
└─────────────────┘

 ┌──────────────┐
 │ Signal Control │
 └──────────────┘

  ┌──────────┐
  │ Physical │
  └──────────┘
```

Figure 2-1. The layered approach.

of functionality is that concerned with the control of the signals, in the form of data, which pass over the transmission media. This process is concerned with who sends and who receives, who "speaks" first and who "listens." The Network Manager who does not understand this element of the specification is creating many problems for the future. Universal connectivity, as is prevalent on the public switched telephone network, does not exist for the private data network.

Another layer is that of data preparation. Once the subscriber identifies the data to pass over the network, the data must be broken into usable chunks (sometimes called packets or datagrams) for manipulation and control. This also includes who or what decides which routes through the network the data will take. For example, this occurs when a subscriber passes a file to a print server for printing. The command from the command line, or the selection of an icon representing that printer, directs the network to route the file to a given location corresponding to the printer's network address.

Finally, at the highest layer, parts of the specifications include the format in which data is to be stored and manipulated. Much of this decision is based on the network operating system software involved and the application software that runs on the various computers and servers attached to the network.

Path Approach

The layered approach as shown by the OSI model is not the only way to begin. The Network Manager can also use an approach based on paths of data flow, as shown in Figure 2-2.

Note that this process is based on physical locations and the software in use. Like other networks, it has a lowest common denominator, in this instance, the telecommunications protocol in use. There are

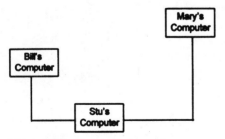

Figure 2-2. The path approach.

several of these protocols, both proprietary and open. This approach views the network as a transport device only, somewhat like a highway, with entrances and exits positioned where necessary.

The Network Manager draws virtual paths that the data takes from one subscriber to another. Common paths are identified and separate subnetworks are created, based on these common paths. This approach is more common when many users perform a common function, such as all employees in one division (accounts payable or R&D). From these separate paths, the Network Manager installs devices that connect these paths in ways which are logical and cost effective. These interconnect devices are either repeaters, bridges, routers, or gateways, depending upon a number of variables that will be addressed later. One key point to keep in mind is to not become bound up with the words; the concept is what is important at this time. The Network Manager must be concerned with what the piece of hardware does, not what it is named.

We have looked at two approaches, the layered approach and the path approach to network specification. These are not the only ones that can be used, nor should they be thought of as such. These are methods that have been found useful in the past for most organizations using data communications networks. The Network Manager's organization, however, may not find either of these acceptable or even workable. Circumstances alter cases and the Network Manager is the one who will be held responsible for the decision concerning which approach to use. Regardless of what approach the Network Manager uses in the specification phase, objective data must be obtained first. There are three general sources of that data, which are the subject of the next section. But before getting into that, it should be mentioned that there is one very subtle trap in this process.

By dealing only with functions, and not brand names, the Network Manager is providing for much more functionality. If, for example, the Network Manager finds the layered approach more attractive, a more open system of protocols may be selected. But if the organization's data communications needs are built around International Business Machine's (IBM's) SNA path approach protocol, then the layered approach would be a bad decision. Another, more common example, is that of specifying "Ethernet" as a telecommunications protocol. True Ethernet is a product of Xerox Corporation. What the Network Manager may intend is that the protocol should be a Carrier Sense Multiple Access/Collision Detection protocol. Or, another way to avoid brand names is to specify a standard instead of a proprietary product. For example, instead of saying token ring, an IBM product, specify IEEE 802.5, a standardized version of the token ring protocol.

By avoiding proprietary products, the Network Manager allows interoperability between unlike systems. Let us assume that the TCP/IP protocol suite is selected for information interchange between like and unlike systems. Second, let us assume that the information exchanged is an ASCII file. By selecting such a nonproprietary approach, the Network Manager assures that departments using MacIntosh computers can share data with departments using IBM hardware. Both vendors offer TCP/IP software to ensure such interoperability.

Sources of Data

There are two sources of data commonly available: the "bottom-up" method and the use of internal questionnaires or surveys. Both methods and their usefulness are addressed below.

Bottom Up

How does this process work? It is common knowledge that the accounts payable (AP) department and, for example, the shipping department, need to share some data concerning the receipt and non-receipt of goods and materials. Therefore, the Network Manager can assume that some type of connectivity will be established between these two departments. This is demonstrated via the path approach. Who else does AP have to communicate with? Assume the organization hires temporary or contract employees. Does AP have to communicate with the human resources (HR) department concerning payment for these nontraditional employees? What is the need for AP and R&D to share data?

The Network Manager should create a bubble chart by drawing a circle for each department and labeling it with the department name. He or she should then draw a line from that department to every other department with which it interacts in the day-to-day operation of business. The final mesh diagram will include all users and elements of the organization.

Once the mesh is complete, the Network Manager must circulate it through each department for each manager's analysis. While it is common sense that AP and HR must share data, the procedure reveals who else AP shares data with that only the AP manager knows about.

One additional strong point flows from this: The subscriber is somewhat involved in the design phase and will not feel as if the new system is being forced upon them from an unconcerned "technoid" who does not understand their responsibilities.

There is a drawback to the bubble chart approach. The Network Manager is painting with too broad a brush. It is all well and good to know that AP and HR share data concerning payment of temporary employees. But how often does this data flow, and how much is there in the flow? This information does not come from the internal structure, but from the subscriber's very personal point of view.

When doing bottom-up specification, as is normally the way a bubble chart is done, the Network Manager must be very concerned with those who do the work. In a true bottom-up specification, each separate department, division, or other element of the organization describes the subnetwork best suited for them. The Network Manager must then integrate these subnetworks into an enterprise-wide final network. This is not as outré as it sounds. In many organizations where the networks were allowed to grow on their own, this is the approach with which the Network Manager is saddled. This is sometimes known as specification by installed base.

There are several very large organizational problems with the bottom-up solution to the specification process. These include motivation, accountability, and cost. Let us examine the motivation problem first.

Is the department or division head who will specify the subnetwork motivated to do so? If this individual has two equal priorities (e.g., next year's budget and the network specification), guess which one will be done first? In many organizations, the Network Manager will be held at fault if the department manager does not complete the design, yet the department manager has no motivation to do so, nor does the Network Manager have any tools to motivate the department manager to complete this task.

We will assume, for the sake of discussion, that the department manager is a nice person, well meaning, and provides the Network Manager with some type of specification for the department's network. Is the specification technically correct? Can the Network Manager assume that it is? No, that would be a poor assumption most cases. The Network Manager will be held accountable if the f design does not work. Yet, how many iterations of the specific and design loop will be completed before the department ma throws up his or her hands in disgust and tells the Network M to build something and let it go at that?

Finally, costing the effort is part of the problem. If the Network Manager's budget is the one which allows for the purchase of the network, then how can department managers get involved? Yet, if a department manager purchases the hardware and software for the department, how can the Network Manager enforce the specification parameters? While an authoritarian management can force compliance, the final product will lack the functionality, and more importantly, the acceptance that a mutually agreed upon specification can offer. Is there a common ground here, and if so, what is it?

Case Study

The organization is a small, fewer than 60-person service bureau, providing value-added services to a commercially available product. The organization has three distinct, but unofficial, levels of the grade structure. The workers are treated as fungible goods; they can be trained quickly and released even more quickly if they do not perform. The owners are absent with other entrepreneurial activities and depend upon a middle level of management to handle the day-to-day operations and short-term planning.

We were retained to provide local and wide area networking skills so the firm can leverage its investment in hardware and software associated with both the networks in place and the ones to be procured. As part of our initial instructions, it is made very clear by the owners that no contact is to be made with the workers and that middle management will tell us everything we need to know. Operational details are to be garnered from these people.

Lesson Learned

As a result of this arbitrary elimination of input from the people who do the work daily, the network as designed runs slowly, the applications are awkward to use, and the service provided does not increase in either quantity or quality. Once the owners found there was no gain in production, they asked why. We provide them with documented evidence that the knowledge held by middle management was not the true way in which the firm did business. As usual, middle management is more interested in the bottom line, as mandated from the owners, rather than true productivity.

Preventive Activities

nce we did get together with the workers, the network supporting applications were redesigned, mainly through segmenting. speed of response increases dramatically, making the customers much happier with the reduction in hold time. Minor redesigns in the applications eliminate double data entry along with its concomitant errors. Accuracy and speed of response increases threefold.

While the bottom-up or bubble-chart approach is one possible method, it may not be sufficient unto itself. The Network Manager must also consider another way to get information from the source.

Survey and Questionnaire

Survey and questionnaire processes are well known to many organizations. Where does information on customers' needs come from? Much of this customer need data come from the field sales force, those people who are in touch with customers on a day-to-day basis. Another source of information is one of more immediate impact, the soldier in the front lines. He knows which way the bullets are coming from and, if he survives long enough, he can pass that information back to his superiors. The same model applies to network specification and design. The person who uses the network on a daily basis is the best one to specify the network. Not on a detailed level, but on a general functional approach, supported by written information. This type of data is gained through questionnaires or surveys. A sample of a useful but incomplete survey is included in Appendix A. It is incomplete because any survey must be customized by the Network Manager to a given organization. There are several points to consider when creating, giving, and analyzing a survey.

The results of the survey must answer certain questions: Note the word "results." The survey or questionnaire is nothing more than a list of questions. The answers to these questions provide data the Network Manager needs to answer specification questions. Therefore the first step is to determine what the specification questions will be. Some representative questions are as follows.

1. What type of communications is in use?
   a. PC to PC
   b. Terminal to mainframe
   c. Client-server
2. What hours is the network used?
3. What are the peak usage hours?
4. How large are the files transmitted and received?
5. How often are files transmitted and received?

6. Do seasonal peaks and valleys occur?
7. How many people do work of type X?

Analysis of Data

The single most important question the Network Manager must answer is "How much bandwidth is needed?" The answer to that question is measured in bits per second (bps) or multiples thereof. Notice the unit of measurement is bits, not bytes. There are eight bits per byte. Most operating systems give file sizes in bytes.

What other questions does the survey answer? The subscribers use specific types of software in their day-to-day operation. The survey will show how many copies of the software package are needed, so the company will purchase either a sufficient number of legal copies of the program or a site license, depending upon which is offered and which is less expensive. The organization is thus kept out of legal trouble. Another question answered is how many employees have access to sensitive, confidential or secret information? Based on this, is there a need for a separate network for this information? If not, is there a need to prevent leakage of this information? What about mission critical functions? If a computer is controlling an assembly line, what would be the impact of network failure between the computer and the assembly line? Which computer(s) are used in areas that have impact on life support or offer danger to human life? What type of backup networks are needed if primary networks fail? The answers to these questions determines the structure and wording of the items to be part of the Network Manager's survey.

Administration of the survey is, in itself, time consuming and important. Remember, the data which come from this survey are used to specify the new network. This specification will drive the design of the new network. And the design of the network will reflect in the career, or lack thereof, of the Network Manager. So, in the administration of the survey, the Network Manager should make much effort to ensure the answers are complete and correct. It is not the purpose of this book to get into questions that check for the survey's internal accuracy. That type of information can be ascertained through other textbooks or from those who create surveys for a living. Still, the Network Manager must realize that surveys must have some means of checking for internal accuracy. The best way to do this is to ask the same question in two different ways. If the respondent provides the same answer to both questions, then the Network Manager can be somewhat assured the rest of the answers are well thought out.

Who gets a chance to respond to this survey? Depending upon the size of the organization and the size of the Network Manager's staff, the answer can range from a small percentage to everyone in the firm. At a minimum, at least one person from each department, division, or other easily defined segment of the organization should have the chance to respond. The responses can be anonymous or not, depending upon the subscriber's wishes. The Network Manager may want to have a statement such as the following somewhere in the survey.

It is not necessary to put your name on this survey. If you do, we may get back to you with follow-up questions. Please indicate if you would like to become a member of the Network Users Group in an advisory or information-gathering capacity by completing the statement below.

I would like to be part of the Network Users Group_____(Y or N) My name and extension number is:

Name_____Telephone #_____

Those subscribers who respond in the affirmative should be followed up with some type of meeting or further survey to determine the depth and breadth of their knowledge and their commitment to assisting the Network Manager in the area of networks.

When designing the survey, the Network Manager must remember that the survey answers must be quantifiable. Answers should be given as numbers, multiple choice, Yes/No, True/False, or on a range from zero to five, or some other fixed, whole number. Questions that are answered as "most of the time" or "once in a while" cannot be quantified and subjected to mathematical analysis. If a subscriber says they use a modem once a day, how does the Network Manager determine the size of the modem pool? If, on the other hand, the subscriber says they use a modem 22 minutes a day, then the Network Manager can, with some degree of accuracy, determine how many modems will be required for X number of minutes per day, X being the sum of the subscribers' answers.

Review

This section has discussed the mechanics of gathering the data which will be used for analysis. There are two sources, the bottom-up input received from department managers and the survey approach. The following section describes the procedures used to determine some specifications from survey responses and other organizational data shown in Appendix A.

Rough Outline of Requirements

This section will take the Network Manager through the steps of an analysis of the data gathered in the survey and other data captured through previous efforts.

1. The data analysis will provide four different databases
   a. Subscriber database
   b. Frequency of electronic-based communications
   c. Frequency of paper-based communications
   d. Software distribution database
2. The subscriber database will list the following:
   a. Subscriber name
   b. Division
   c. Department
   d. Office
   e. Type of computer in use
   f. Type of terminal in use
   g. Modem usage
   h. Facsimile usage
3. The electronic-based communications database lists the frequency and size of communications with other
   a. Divisions
   b. Departments
   c. Offices
4. The software distribution database lists software in two categories, that which is currently in use and that which the subscriber identifies as being necessary for future efforts. The information includes
   a. Subscriber name
   b. Division
   c. Department
   d. Office
   e. Software for current usage, e.g., word processing
   f. Software identified for future usage, e.g., statistics and CASE tools

The Network Manager should start the process by creating a spreadsheet as is shown in Figure A-1 in Appendix A. This spreadsheet shows a small time slice of existing and potential traffic received by each employee. By summing all received traffic and converting the pages to bits, as explained in the Appendix, the Network Manager can determine bandwidth required by each employee. This could have been done with transmitted data and paper as well. The choice of whether to identify transmitted or received information is the

Network Manager's decision. The resulting information is then entered into the appropriate databases.

Now all the Network Manager must do to determine a requirement is to query across one or all the databases by one of the key fields.

1. Subscriber name

2. Division

3. Department

4. Office

Then to this query, add what information is desired from other fields. Sample queries might answer such questions as "How many copies of statistics software packages are in use by the R&D department? Which department is using them today and how many copies will we need later? Why should these questions be asked? The Network Manager must be able to identify current software needs as well as forecast needs for the future. These new needs will put a load on the file servers if a client-server model is in use, and will put larger demands on backbone circuits regardless of whether the client-server model is in use or whether BMC uses the peer-to-peer model.

Another sample query to use is What is the total amount of bandwidth being consumed by communications between accounts payable and personnel, by time, over the normal work day? This information will allow the Network Manager to determine whether additional circuits are necessary or whether spare circuits would be financially justifiable.

In general, the analysis of data that have been captured is controlled by the needs of BMC's network. The questions are many and varied and well beyond the scope of this book. The Network Manager is the one who must consider the questions, phrase them correctly, and conduct queries across the databases for answers. This process is neither simple, easy, nor quick. Much effort and thought must go into the process because most of what comes out of these queries will be used to describe the network to be designed, installed, and tested.

Rough Outline of Requirements

The result of the Network Manager's data gathering is shown in the spreadsheets or databases. We know how much throughput is required between subscribers. We have identified the major sinks and sources of data. We know, to a certain extent, the distribution of existing network assets. And finally, we have a rough outline of network requirements. This provides us with a quantitative analysis that is, by itself, insufficient. The Network Manager must be more than a bean counter to be successful in the specification phase. The following two sections explore the nonquantifiable elements of the specifications.

Identify Critical Mission Requirements

Any organization has critical mission requirements that must be addressed in the specification phase. The mission criticality will vary depending upon the organization, its goals, culture, and environment. There are three aspects of satisfying critical mission requirements: single point of failure, redundancy, and alternate routing.

Single Point of Failure

Figure 2-3 shows one possible design based on specifications derived from the data gathered and shown in Appendix A. What is wrong with this design?

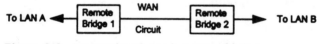

Figure 2-3. A network with single points of failure.

Note the connectivity between remote Bridges 1 and 2. This transmission media and the equipment connected to each end of it are single points of failure. That is, if the equipment identified as Bridge 1 fails, all communications through that device stop. Subscribers on each side of the device may, under some conditions, pass data to each other, but not across the device itself. The same can be said if the transmission media fails. There are two generic ways of bypassing the single point of failure problem: redundancy and alternate routing. The less attractive approach, redundancy, will be examined first.

Redundancy. This is the duplication or triplication of hardware in support of mission-critical communications requirements. The only good thing to say about redundancy is that it is cheaper than alternate routing. In the end, the decision of what to do is based on the cost/benefit ratio of the solution.

To take an example, we will assume Variable Reality (VR), Inc.'s sales force has a reputation for making Just In Time buyers look like heroes. This is accomplished by technology both with the field sales force and VR's in-house sales force. The field sales force can, while at the buyer's location, use their laptop and modem to dial into VR's internal network, identify the quantity of parts on hand, and estimate their delivery date to the buyer's factory. When the salesperson dials in, he or she connects to a device we will call a black box. The hardware and software inside the black box makes the salesman appear to be just another node on the network. Now, consider the impact of this black box dying in the middle of a query concerning the number of products ready to ship. If redundancy is the method used to eliminate a single point of failure, the cost/benefit ratio is easily computed.

Let us assume the cost of the black box is $2100. Data gathered in the survey show that it is used only 128 minutes in any given eight-hour day. Therefore, only one box will be needed. Further, let us assume that the minimum profit from any sale is $80, and the maximum profit from any sale is $900. Based on these two extremes, we have an average profit per sales call of $980/2 or $490. These figures are arbitrary and may not reflect reality. But, assuming they do, it is easy to compute the cost of lost sales based on one defective black box. For every lost sale, the profit goes down by $490. How many sales that are not lost will make up for the expense of a second black box? This is fairly easy to calculate: Divide $2100 by $490 and the result is 4.29. Therefore, we find that five (4.29 rounded up) sales which were lost due to a defective black box will pay for a redundant black box.

This type of redundancy is easy to analyze and plan for at this stage. But what about redundancy at the file server level? Should there be two file servers for each one identified in the specification? It is certainly easy enough to justify such a duplication financially. Consider the cost if the entire shipping department has no computer connectivity for one day. Or what happens if the R&D people cannot do the intensive number crunching, which is part of their everyday activities? How great is this loss in terms of dollars or competitiveness?

Anyone who is good with numbers can manipulate figures to show that each department in the organization should have a fully redundant network, with both systems running in place at the same time, to ensure 100 percent availability. In counterpoint, anyone who is senior in management also knows how to manipulate the same numbers to show a different result, having done so before. This impasse is not as great as it may appear on the surface.

Many of the components used in modern electronics can trace their heritage to those used in military and aerospace programs, both operational areas requiring very robust components and long mean times between failures. With modern electronics, we have found that if the device does not fail within the first 30 days of operation, it will not fail for many years. So, perhaps the only redundancy required is for mechanical devices such as hard drives, printers, or scanners, where there is mechanical movement. This is not to say that redundancy is not necessary—redundancy is required in some hardware. The choice of what hardware to make redundant depends upon the perceived loss to the organization when the hardware fails and the Network Manager's skills and experience in the area of influencing such perception.

Alternate Routing. On the other hand, alternate routing (alt route) is more favorable as a means of ensuring that critical mission requirements are achieved. Alternate routing partakes of some elements of redundancy in as much as it duplicates certain hardware. However, the hardware that is duplicated is not usually expensive; the expense with alternate routing is a monthly, recurring expense of unused transmission media. This falls into two categories; LAN media and wide area network (WAN) circuits. There is very little in the line of recurring expenses with LAN transmission media as it is usually owned by the organization that owns the LAN. Such is not the case with WAN circuits, which we will discuss after a few comments on LAN transmission media.

LAN transmission media used for alternate routing are difficult to design. When a building or campus is wired for the LAN, it is done so in the most cost-effective, intelligent way, thus ensuring that transmission media is routed in the easiest, most direct path possible. By doing so, the installers often use existing tunnels, shafts, air spaces, ducts, or conduits already in place. So will others who come after them. Those who follow after the LAN installers can do much damage to relatively fragile fiber-optic or twisted-pair cable. Cable that is run under the floor is subject to jack hammers, cold chisels, and hacksaws. Cable that runs in tunnels is subject to being crushed from material being carried through the tunnels, or from flooding or tunnel collapse. Cable that runs in air spaces is more expensive than others and is moved around every time the ceiling is opened up. Someone running a screw up through a ceiling tile can pierce the backbone cable, thereby shutting down the LAN. The best way to prevent damage to LAN transmission media is to run a secondary path from device to device at right angles to the primary path. That is, if a run of fiber-optic cable goes horizontal, duplicate it vertically. If this is not feasible, and in many instances it is not, place a secondary parallel run as far away from the primary run as possible. If runs cannot be placed in parallel or at right angles, select another transmission media. In other words, if a secondary run of copper cable is not permissible, put in a backup transmission path of microwave or satellite linkage. The key point here is that the backup media is in place only to support mission critical requirements, not the day-to-day operational requirements. That is a different story.

With WAN circuits, many of the same principles are true, but their application is much more difficult. Let us assume, for the sake of discussion, that the Network Manager has contracted with the regional or local telephone operating company for a T-1 circuit from one location to another. Being intelligent and somewhat fat in the budget, the Network Manager thinks a backup T-1 circuit will prevent a WAN disaster. Wrong! When purchasing a backup circuit from a public carrier, the Network Manager must ensure two things.

First, it must be determined that the circuit purchased for alternate routing does not traverse the same route to the local point of presence (POP) as does the primary circuit. This includes connection to the nearest switch. Second, and although it is difficult to arrange, it may be useful if the alternate routing circuit is purchased from another common carrier or regional operating company. The Network Manager must be careful that the alternate common carrier's routing does not use the same cables the primary carrier's routing does, however. This is quite common where high bandwidth, fiber-optic cable is used. Where it can be cost justified, and when it is physically possible, it is strongly recommended that the Network Manager consider alternate transmission media. If wire is in use to the POP, the Network Manager should consider microwave as a backup. If geography prohibits microwave, then perhaps satellite may be considered.

There are also three other lesser points to consider when contemplating alternate routing. First, an alternate route provides the foresighted Network Manager a secret reserve of bandwidth on demand. Such quick reaction capability endears the Network Manager to senior management. Besides, it is another argument to draw upon when lobbying for additional budgetary largesse. Second, use of alternate routing as an emergency source of bandwidth on demand is a two-edged sword. It is very difficult, for instance, to take bandwidth away from subscribers used to having it. Therefore the use of alternate routing bandwidth is discouraged except for outages of the primary path. Finally, the Network Manager should consider configuring an alternate route at less than the bandwidth required for the normal circuit. The Network Manager pays for bandwidth by distance and the amount of time the circuit is in use. As the alternate routing bandwidth will not be used except to support critical mission requirements, the Network Manager does not need to specify as much bandwidth as is required for all daily operations.

One final point of consideration for review here. When—not if—disaster strikes, should the switchover to the alternate route or redundant equipment be done automatically or manually? There are pluses and minuses on both sides of the argument. Automatic switchover eliminates the need for human intervention. This is good if the staff is small or unskilled, or the organization uses "lights out" networking and computing. This is bad, however, if the problem is momentary in nature or switch back to the main circuit or hardware is automatic as well. Manual switchover is slow and requires some skilled intervention. Still, it can be positive in nature because it is under constant control. A hacker or recently terminated and revenge-oriented employee cannot get in and trip the switchover mechanism. The manual approach is bad, however, because it requires skilled people to be on site at all times to facilitate the switchover and switch back.

Review

This section has presented numerous points and opinions concerning the procedures and requirements for ensuring critical mission requirements that are considered in the specification phase. Much of this information comes from data gathered through surveys and from bottom-up specifications. But none of the above can substitute for the Network Manager's knowledge of the organization.

What Is the Business?

There are two areas of concern at this stage of the specification process: how the changed network will support the business of the Network Manager's organization and how daily operations will be modified by this changed network.

How Will the Network Support
the Organization?

Many, many organizations make the same mistake. They use the network to eliminate the paper which shuffles through the mail room. So, all that has been accomplished is the loss of one or two positions which could be filled by entry-level, unskilled workers. The organization has spent hundreds of thousands of dollars to eliminate a $5.00 per hour courier. Progress, thy name is mud!

To see one of the more common ways in which a network changes the way a firm does business, walk into the nearest shopping center or mall. Look around, find the nearest ATM. How much would it cost a bank to put a teller in each shopping center and mall, for 16 hours a day, seven days a week for minor transactions such as withdrawals and deposits? Considerably more than it costs for the ATM machines and their installation. How have ATM machines changed the way the bank does business? There are many ways to demonstrate the changes.

First, the bank needs fewer tellers at even fewer branch offices and can keep these offices open shorter hours. This reduces labor costs. The way in which a manager computes the return on investment (ROI) for such an installation differs from bank to bank. Still, it is safe to assume that reduced labor, with its concomitant reduction in expenses will be a major factor in the calculations.

Another example of how a network changes the way an organization does business is the efforts of parcel delivery companies. Each shipping label has a unique number. The old tracking method was for the driver to note on a form when and to whom a parcel was delivered. This information was manually entered, or at best scanned in, the end of each day. Now the driver waves a bar-code reader over the shipping label, which stores the date and delivery time in a field service data storage device. Next the name of the recipient is entered into the same device, and this information is downloaded at the end of the driver's shift. The shipping company can determine the location of a given parcel at any time. It can be determined to be either in the shipping point awaiting pickup, in transit to the final location, in the vehicle being delivered to the recipient, or already delivered to the appropriate address.

The past two examples may or may not apply to the Network Manager's organization. Let's face it, aside from automating the assembly line in a manufacturing firm there may not be a lot a network can do regarding the way in which a given organization does business. Office automation is a foregone conclusion; how can that affect the way the firm does business?

How Will Daily Operations Change?

Changing the flow of paper is one possible way of affecting daily operations. But when this happens, we find many middle managers, who had previously had been responsible for data gathering and analysis, now find themselves doing less and less work. Soon management will note this, and reduce their positions accordingly. This brings us to the second point, how the daily operations will change.

Before. Let us assume Sally Green, an engineer in a company's R&D department, needs a voltmeter for a new project. She gets verbal approval from her manager. She then creates a purchase order and routes it to the purchasing department for its blessing. Once this is completed, she contacts the vendor and orders the meter. It is received on the loading dock, then routed to her lab. The loading dock people have signed for it at delivery, acknowledging receipt. This form is routed back to the accounts payable (AP) department for approval and payment when the bill is tendered. The check is cut and mailed. Now, we will network this process.

After. Sally sits at her PC and generates an e-mail message to her manager describing what she wants and how much it costs. Her manager appends an electronic approval, forwarding a copy of the approval to purchasing and back to Sally through e-mail. Purchasing does a cut and paste of the name of the vendor and product specification for the purchase order. The clerk in purchasing, Joe Smith, completes the purchase order by assigning a purchase order number available from AP. He obtains this number by accessing a protected list of numbers using his password. At the same time, he enters the cost and product ordered into AP's file so that their records correspond to this number. Joe enters the same order information into a database under due out files. Joe sends the order to the vendor by facsimile from his personal computer via a facsimile-capable modem.

The vendor boxes the item and ships it. It is received on the loading dock at Sally's firm. Mike Jones in the receiving department notes the purchase order number and opens the due out file. Mike enters the date and time of receipt, thereby closing the particular record identified by the purchase order number. The voltmeter is delivered to Sally, she informs her manager via e-mail that the delivery is complete, and then goes about her job.

Several days or weeks later AP gets an invoice from the vendor. They review the database, noting that the meter was delivered on a given date and received by Mike Jones. AP authorizes a check to be cut for payment for the voltmeter. The check is issued and mailed to the vendor.

This is one very simple means of using the new network to affect the way in which the organization does business. Is there a more complex way, or a method which has a greater impact on the organization? The Network Manager must be the one to find out. In doing so, the Network Manager also must include specifications to solve the problems associated with growth in the organization. The next section addresses that particular effort.

Growth of the Organization

No organization is static for long. Changing market share, in(de)creased revenues, regulatory changes, and the fickle public cause the organization to grow, shrink, or otherwise change its structure. The change could be outward, with an increase in the number of employees; it could be upward, with more and more decision making centralized at the top; or it could be downward, with decision making pushed down the structure to the lowest possible level. How is the new network specified in relation to these ongoing changes? Let us look at each of the three directions just mentioned: outward, upward, or downward.

Outward change merely means more of the same activity. The Network Manager must allow for increased bandwidth, perhaps at new locations outside the immediate campus or geographic area. The Network Manager must also plan for more hardware. This may mean that the database used for configuration management may have to be scrapped and replaced by a more powerful one. Or the Network Manager may have to hire, train, promote, and reassign more technically qualified personnel at the new locations. Another problem with outward growth is in assigning subnets and subnet address blocks. Addressing will be part of Chapter 3 on design, which follows.

If the organization changes upward, the Network Manager must modify the specifications accordingly. Those at the top will be fed much raw data as reports, analyses, and recommendations. This means that the transmission media going to these people must have much bandwidth and be quite reliable. The hardware in use at the upper levels must also be powerful in its number-crunching capabilities. This means additional training for the users who, in all likelihood, are not well versed in many of the more common applications used to analyze and manipulate data. Here the problem is one of bandwidth and topology.

If the organization changes downward, with decision makers situated low in the organization, then many, many circuits are being used of lower bandwidth than the circuits used for the upward mode. Here redundancy can be used in place of the high reliability circuits that are used in the upward mode. The major problem with this approach is the use of different applications. The AP department may use one vendor's word processing package, while the personnel department uses another, incompatible one. Here, the problem is one of specifying the integrated whole, not just resolving the bandwidth and topology problem.

Growth is not limited only to the immediate future, but affects long-term changes in areas such as protocols and topologies. The venerable X.25 protocol has changed every four years, yet people still use it to pump data through networks of limited bandwidth. The Network Manager must look at protocols and how they have changed through the years. With this historical perspective in mind, the Network Manager must forecast where these and other protocols will be going during the allotted lifetime of the specified network. The new network must be specified to support these current and future protocols and topologies. At this point, the network specification is reasonably complete from a bandwidth, hardware, and software viewpoint. The Network Manager has one final set of specifications to produce, but they are of concern to a very limited set of people—those on the Network Manager's staff. This decision is the network management system to be used to run the network.

Network Management System

The network management system specified must be able to perform tasks within the five standard functional management areas (SFMAs) identified by the International Standards Organization. These include

1. Accounting management

2. Configuration management

3. Fault management

4. Performance management

5. Security management

We have detailed many of the aspects of these five SFMAs in our companion book on the subject of network management. Therefore, we will touch on each of these areas only briefly.

Accounting management is concerned with who is consuming what services and the bandwidth available on the network. Configuration management deals with what is where and what version, serial number, address, or other data that concern a particular device or software. Fault management encompasses tracking and eliminating error conditions. Performance management is concerned with ensuring what the organization purchased in performance is what is being delivered. Security management is arranged around the integrity of the network and the data which flows through it. This network management system supports the Network Manager in three general ways: by means of reports, guidance in the day-to-day operations, and response to direction from the Network Manager or staff. Each of these methods are detailed below.

What Reports Are Needed

One of the most common cliches of management is "If you can't measure it, you can't manage it." Measurement begins with data gathering. Data gathering finishes with a report in a format that is easily understood. What reports are needed to facilitate management of the network? The total number and format will vary from organization to organization, but the following list includes several.

- *Bandwidth consumption.* How much bandwidth is in use at any given instant?

- *Addresses assigned.* Which device has which electronic address, and where is it?

- *Trouble tickets.* How many are outstanding and how many have been completed in time X?

- *Password change.* When was the last time subscriber Green changed passwords?

- *Circuit cost.* How much did it cost for leased circuit 123XAC-44567 last quarter?

- *Outages.* How long was bridge 4N18 out during the last year?

- *Authorization.* Which subscriber(s) is authorized connectivity to which device(s)?

Some reports are more important than others. Some network management systems provide reports in graphic formats, normally a moving curve or skyline depicting the consumption of bandwidth or another dynamic process. This can be particularly effective as a demonstration to senior management of the high-tech stuff their money has purchased. Whereas this is not necessary for effective network management, it is effective for management of those who have the finances.

Some reports are needed on a minute-by-minute basis. A bandwidth consumption report is one of these. Others are needed daily, such as, perhaps, a report indicating the number of trouble tickets opened and closed on a given day. Other reports are needed monthly, such as circuit cost reports to verify billings. Still other reports are needed on a nonroutine basis. For example, there is no need to change an existing report on addresses if the addresses have not changed; yet this report must be generated when any given address does change. Authorization reports must be generated when new employees are authorized access to a device, or when these same employees leave.

Network Management Systems Expertise

Network management systems come in various levels of expertise. Some of these are no more than a statistics gathering program which checks measurement points on a recurring basis. The program then stores the raw data to a database where the Network Manager can extract, manipulate, and evaluate the numbers.

Other network management systems are identified as expert systems or automated network management systems. These systems perform data gathering, analysis, and report preparation tasks. They also switch defective equipment out of a circuit and switch hot standby equipment into the circuit, without human intervention. The system then generates a trouble ticket, informing the technician which piece of equipment is defective.

Other forms of expert systems perform troubleshooting when coached by a human user. This type of system offers a choice of potential defective elements on the network; a human being uses his or her intelligence and experience to eliminate those components which do not have a strong possibility of causing the problem. As with most things, the flexibility and applicability of network management systems is directly related to its cost. Some of the better offerings from major vendors can run in the mid to high five figures.

Responsiveness

The network management system must be responsive in several areas, the most important of which is speed. If troubleshooting information is needed now, it must be presented now, not five minutes later when the network management system has had a chance to manipulate the data. This is a function of two things: the processing power of the CPU used in the network management system and the bandwidth consumed in gathering data. The faster the CPU, the faster the data can be manipulated. The greater the bandwidth, the more data can be obtained in a given time—up to some physical limit. We can think of very little need for a T-1 circuit dedicated to network performance data gathering.

A less important area of responsiveness is that of accuracy. If part of the network management system is to evaluate the physical quality of the network, it may include a time domain reflectometer to measure the cable itself. How accurate, in feet or meters, is the measurement provided by this time domain reflectometer? If the network management system measures throughput, accuracy may be concerned with more than the number of raw bits flowing through the network. It may be concerned with the quality of the packets which consume bandwidth, or the exact size of these packets.

A final measure of responsiveness is that of report customization. A report which is unusable is a waste of paper or disk space. If the network management system works with both IEEE 802.5 and IEEE 802.3 based networks, but the only protocol in use is IEEE 802.5, the reports concerning IEEE 802.3 specifications may be unnecessary. Can the network management system be customized to eliminate, or at least suppress, undesired and unused information?

In summary, the network management system must provide the Network Manager with reports, it may provide certain levels of technical expertise based on the Network Manager's budget, and it must be responsive. By specifying which elements to have available in such systems, the Network Manager has almost completed the specification phase.

Review Phase

There is no set format for the specification stage as there is for the RFQ or contract. Most of the specification exists on scratch notes, e.g., " . . . bridge 2B4 to 2B5—bandwidth 256 Kbps." Or "Keep all networks 802.3 compatible." No, the specification which must be reviewed here is one of functionality and connectivity. That does not mean, however, that documentation is not required. But first, let us look at the questions to be asked, and who will answer them.

Types of Questions for Subscribers

These are functional questions asked to everyone from the most junior analyst to the most senior manager.

Do you share spreadsheet data with_____?

How often do you connect to_____?

After the questions are answered, the Network Manager must then say something like, "I'll be providing connectivity and information interchange between you and (fill in organization names here) so you may (fill in function here). Any last minute changes you'd like to see incorporated in the new network?"

Based on the answers to these questions, the Network Manager must review the specification for fulfilling these needs. This may seem to be an iteration of the survey process performed earlier, but it is not an iteration; it is a final check to see if anything has changed. Depending upon the duration of the Network Manager's plan for this effort, the specification phase can take anywhere from two weeks to two years. Departments may be created, others eliminated, some grow, and others shrink.

Who to Review With

Various people other than the subscribers should be contacted for questions at this point. The Network Manager should review the specifications with the internal telecommunications or MIS staff for their last minute changes. Too often people say "Oops, I meant to add (fill in the component here) to the final spec and forgot it. Sorry 'bout that." If the Network Manager has a good idea of which vendors will be selected as suppliers and installers, this may be a good time for a last minute review with them. They may, however, be too busy or unwilling to commit free time for a nonpaying effort: That is the way of the world, and the Network Manager must realize this. If free help is not available, the Network Manager may wish to purchase the time of a consultant for the final analysis of the specification phase. And, after everything else has been double-checked, the Network Manager must ask the most important question of all: "What have I forgotten?"

Practice Defensive Management

In most organizations, most employees willingly support the team effort for the achievement of the organization's goals. But since some do not, the Network Manager must prepare a defense against those who will decry the specification effort as poor and unrelated to the way the organization does business. Much of this hue and cry comes from ego; persons who were not consulted for their input in the specification phase, or who will lose the right to use their favorite network or application in the new installation. The Network Manager can defend against these people in the following manner.

He or she should prepare a memorandum indicating that the specification phase is nearing completion. The memorandum offers everyone a chance to provide final input into the finished product, but only in a written format. The memorandum must state a date by which the final input is to be received. If the potential complainers do not return the information by that date, it is not the fault of the Network Manager and he or she may wash their hands of any problem arising from these non-team players. If they do return a response, then the Network Manager can say it was reviewed and incorporated where it seemed useful. In other words, the Network Manager has two possible answers to any complaints received:

> You were given a chance to reply, the cutoff date was noted, and you failed to meet the cutoff date. My schedule could not be allowed to slip because of your tardiness.

Or,

> You were given a chance to reply and did. We analyzed your input and selected those items which seemed to fit in with the overall scheme of things. However, you must realize other departments have needs which, if your recommendation was accepted, would prohibit them from accomplishing their portion of the overall goal. That is not to say your reply was wrong or unacceptable. It is just that your requirements were not capable of being implemented for the greater good of our organization.

Summary

In summary the Network Manager must, after completing the specifications, review them with others in the organization. The result of the specification activity is a design. But before we get to the design phase, let us review the contents of the specification phase.

1. Specification starts with the identification of the functions desired.

2. Functions have two sources: a bottom-up approach or question-naires/surveys.

3. The Network Manager must analyze the data to provide some definition of what is needed where.

4. All mission critical requirements are identified and provided for in the specification phase.

5. The Network Manager must clearly understand the nature of the organization and its business.

6. The specification must allow for growth in the organization.

7. The specification must include some type of network management functionality.

8. The Network Manager must ensure a final review is completed before beginning the design phase.

3
Design

As the purpose of the specification was a design, the purpose of the design is creation and formalization of information necessary for the request for quotation (RFQ), sometimes known as request for procurement (RFP). For this book the terms RFQ and RFP are considered interchangeable.

In previous chapters we have mentioned the principle that a small amount of effort expended in the beginning of a job will reap large rewards at the end. This leverage approach changes as the Network Manager steps through the procurement process. The design phase has some impact on the final product; that is a given. Yet all the effort in the design phase can be nullified by a poor effort in the RFQ or in the testing procedures. At this point in the network procurement process, the Network Manager must realize that changes which occur after the design phase is complete are extremely difficult to implement and their impact is quite significant in terms of increases in budgeted expenses and the appearance of nonbudgeted expenses.

As there are several steps in the design process, this chapter will examine each of them in the recommended sequence in which they are to be completed. The Network Manager must

- Analyze by throughput
- Determine physical distances

- Rank order topologies
- Pick transmission media
- Select hardware
- Create the test plan

One of the few things that all networks have in common is the design which, in turn, is driven by throughput in terms of bits per second or multiples thereof. It is only natural that the Network Manager's analysis begins with this.

Analysis by Throughput

The analysis is nothing more than an ordering, in ascending or descending sequence, of what sinks and sources of data transmit and receive data in any given instant. The Network Manager should, as the review of sinks and sources begins, realize that traffic flow is not intuitively obvious. Someone who is knowledgeable about computer communications may think that a mainframe is a large source of data traffic and should have priority of access to that part of the network with the greatest throughput. This may not be the case. Let us again examine the example of the ATM and its impact on the way in which a bank does business and how this affects the design phase.

The Local Last National Bank & Trust (LLNB&T) has multiple ATMs scattered throughout the state. These are connected to a very large mainframe computer from Just Computers & Networks (JCN) which, in turn, holds all the bank's depositors' records. ATM traffic is not truly random, but it is restricted pretty much between seven in the morning until roughly ten in the evening. With this in mind, how much bandwidth on the bank's network is needed?

The first linkage is between the remote ATMs and the JCN computer at the corporate headquarters. Is each one of the ATM machines assigned a dedicated port on the mainframe? No, that is not practical. On the other hand, does each ATM machine need a dedicated, leased line to the same corporate headquarters? Again, the answer is no. Each source of traffic (an ATM in this instance, or subscriber in another network) generates an average amount of traffic. This average will vary from time to time and place to place, but at least it gives the Network Manager a figure to point to and say "This is what I'll build to."

Since we have stated that the ATM does not need a dedicated leased line and the mainframe does not need a dedicated port for each ATM, then what is needed? Dial-up lines and a modem at the ATM and some type of modem pool or front-end processor at the site of the mainframe. Initially the Network Manager will get a regular dial-up circuit from a carrier. The ATM will automatically dial into the headquarters every time a customer wants a transaction. The Network Manager must make an educated guess concerning the number of modems in the modem pool or the number of ports on the front-end processor. It is better to provide too many ports or too many modems than not enough.

As was shown early in the book, the specification and design phases are iterative. This would be a good place to run through another loop of the specification and design to ensure that there are enough ports or modems to support the ATM machines in the field. While some very expensive simulation software may be used to determine the actual number of ports or modems required, we believe this to be a waste of time and money in most organizations. Theoretical approaches merely provide a starting point. Experience with actual usage will provide the data concerning how many ports or modems are needed to ensure the reaction time acceptable to customers. One additional point needs to be made here. As usage data is gathered in the form of telephone bills, the Network Manager may find it is cheaper to provide leased lines for the more active ATMs.

This has been one very brief demonstration of the way in which the analysis starts. We will now present the general approach to design based on specifications, the design by business function.

By Business Function

When analyzing throughput by business function, the Network Manager is preparing to segment the network at a later date. This segmentation, and even subnetting if the process is taken to a sufficient level of detail, will identify the location of bridges, routers, and gateways. But we will be concerned with the groundwork first. The mesh diagrams created in Chapter 2 is a good place to review the procedures to produce a similar diagram.

Start with any department, R&D, for instance. Identify each sink and source of data within the R&D department, including unmanned data gathering devices. Draw a line representing the flow of data within the department from which source to which sink.

Figure 3-1 shows the first step. Here a scientist running a personal computer at his or her desk is making use of the large data base stored in a mainframe computer. The data manipulation is being done in the PC while the mainframe is the data storehouse. All that flows from the

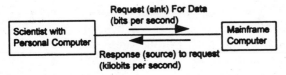

Figure 3-1. Sink to source.

PC is the command to extract data, normally a very small message in terms of bits per second. What flows from the mainframe is a larger message, the data requested, probably in the kilobits per second range. Label the line between the scientist and the mainframe with the throughput, which was gathered in the specification phase.

Who else does the scientist communicate with in the normal business day? The information the scientist has analyzed and the resulting report concerning this data must be prepared for someone; the content and format of the preparation will depend on the needs of the organization. The report may be typed by an assistant. The scientist must forward it in raw form to the computer used by the assistant. The Network Manager draws another line from the scientist to the assistant and labels it with the throughput in bps. What else does the assistant do with the computer on his or her desk?

In many organizations, e-mail is necessary, so the PC may be used for accessing e-mail, both for the assistant and for others in the R&D department who do not have a computer. Draw lines representing the means of accessing the e-mail servers which the PC can access. Let us go back to the report the assistant has just finished. Does the assistant pass correspondence concerning this finished report to others in the organization? Does the marketing department need the R&D information for new product lines? Draw a line connecting the assistant to the marketing department, and one or more of the computers in use there. Label this line with the throughput in bps.

Once all the R&D connectivity and throughput is noted and diagrammed, draw a perimeter around the R&D department, pick another department, and continue with the process. Complete this process for every department and division within the firm. There may be perimeters within perimeters, depending upon the size of the organization. The R&D department may have a chemistry division, an electronics division, and a physics division. Each of these will have their boundaries, but still be considered to be within the R&D department (see Figure 3-2).

Here the Network Manager is presented with a chance to make a very easy mistake. Looking at the diagram, the Network Manager sees the flow of data from all points to all other points, right? Well, yes and

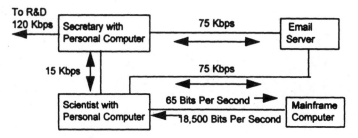

Figure 3-2. Multiple data paths.

no. Yes, in as much as most of the sinks and sources of data are concerned, and no, because not all of them may be there. Yes, because the amount of data flowing between departments and divisions is known, but no because that value may be wrong.

First, we will look at sinks and sources of data which are not accounted for. Will the new network carry bits coming from security devices? Will the future efforts of the organization require the LAN to support video teleconferencing? Are there any "secret" networks in existence that no one other than the users know about? Second, we must make sure that a little addition is done. If the R&D department communications is with the marketing department, the diagram should show two lines. One of these lines was drawn when the R&D department was analyzed, the other when the marketing department was analyzed. Although they represent two different data flows, they will flow through the same transmission media. The Network Manager must ensure that dual flows like this are summed, the sum being the required bandwidth.

Figure 3-3 shows a small portion of the final diagram, which was begun in the analysis of a business function above. Although this is the preferred approach, it is not the only one.

Analysis by Location

The Network Manager can use this approach when departments or other business units are separated by physical distances that cannot be spanned by a single LAN without the use of bridges or repeaters. The Network Manager performs the analysis by business function, then carries the effort further. First, each business unit is analyzed by its internal requirements. The same process is then reiterated on a higher level of detail, the Network Manager being concerned only with infor-

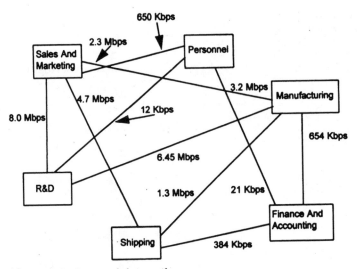

Figure 3-3. Summed data paths.

mation interchange between business units, which are separated by kilometers not meters.

Here the Network Manager can run into a severe miscalculation. Let us say that Joe Smith, the manager for the local plant, is asked to determine how much data flow between him and his immediate supervisor at corporate headquarters, 350 miles away. We are quite sure that Joe will remember the daily e-mail messages, the daily production reports, and probably the weekly summaries. But will Joe remember the monthly database summaries which contain megabits of data? If he does not, and the Network Manager bases the design on small daily and weekly data flows, then the new network will bog down under this unforeseen, transient megabit load. Interlocation network design should be approached with much care; Network Managers must understand that subscribers answering surveys are harried individuals who may miss an occasional usage, with its corresponding throughput.

Another problem the Network Manager must work out in this portion of the design phase is the cost/bandwidth/throughput trade-off. As we have noted, there is a relationship between the bandwidth used and the money available to pay for it. Now we must examine the relationship between bandwidth and subscriber (dis)satisfaction with delay. In a nutshell, the Network Manager must decide how long the subscriber will be willing to stare at an empty screen until the file transfer or response to the query takes place. As a rule of thumb, three seconds is appropriate. If the subscriber moves a 6Mb file from one place to another, the transfer should take no more than three seconds. Six million divided by three provides a throughput of 2 Mbps. This speed is quite common in LANs, and quite uncommon and expensive in WANs. The best the Network Manager can do is to explain to the disgruntled subscriber that money is not available and a slight, occasional delay is the price to pay for this penury. Otherwise, the subscriber could be requested to pay for the higher bandwidth out of his or her budget.

At this point the Network Manager knows how much traffic is flowing from all sources to sinks within the network. This does not mean the Network Manager is ready to start ordering material, but rather, that he or she must do even more design, this part of the design effort being finicky indeed. If the Network Manager fails to follow the rules here, the results will range from no immediate obvious mistakes to a network that simply refuses to function. What are we discussing here? Read on!

Physical Distances Involved

In dealing with networks there are two words that some people misuse and, by doing so, cause themselves much heartache and many problems. These are the terms *topography* and *topology*. A network topology is the type of network: bus, star, or ring. The network topography is the physical location of transmission media in three dimensions. This section will be concerned with the network topography, particularly concerning distances.

LAN Extensions

There are many rules concerning the design of networks, and each must be obeyed or the design will fail during installation. General rules can be ascertained through books on the subject. We recommend Stallings three-volume effort.

Handbook Of Computer Communications, Vol. I, Sams, 1988, 0-672-22664-2.

Handbook Of Computer Communications, Vol. II, Sams, 1988, 0-672-22665-0.

Handbook Of Computer Communications, Vol. III, Sams, 1988, 0-672-22666-9.

Specific rules can only be deduced through discussions with equipment vendors. For this book, we will use the bus as a form of topology, and use a heuristic, collision-sensing, multiple-access, collision-avoid-

Table 3-1. 10Base5
Parameters

Criteria	Value
Data rate	10 Mbps
Segment length	500 meters
Network span	2500 meters
Nodes per segment	100
Node spacing	2.5 meters
Cable diameter	10 mm

Frame (packet) size octets (bytes)	
Minimum	64
Maximum	1518

ance (CSMA/CA) network based on the IEEE 802.3 specification. This is most often confused with Ethernet™, a proprietary product of Xerox Corporation. We also will use the more well-known 10BASE5 version of this specification, with the parameters shown in Table 3-1. This is sometimes known as the DIX standard (Digital, Intel, Xerox).

Based on the parameters in Table 3-1, the following rules apply to the physical limits of the network design. The maximum length of the network is 500 meters or one segment. Yet the network span is 2500 meters. How so? The rules say the maximum number of nodes per segment is 100, using a 500-meter segment limit, each node being separated from another node by 2.5 meters. Dividing 500 by 2.5 meters, the minimum physical separation, provides an answer of 200, which is a direct contradiction to the rule of 100 nodes per segment. What are these rules? Are they the exceptions which prove the rule? No, this is not the case; the rule is that the rules that apply may not all be written down somewhere, waiting the Network Manager's discovery. There are two subtleties here that must be examined.

The first is the 500-meter segment versus the 2500-meter limit. What is not shown in the rules is that a device known as a repeater must be installed between any two 500-meter segments. For those of us who have some exposure to this type of network, the answer is "Oh yes, I should have thought of that." The other question, the node per segment problem, is not quite as obvious. It is related to the definition of the term node, and the address structure of the packet itself. Discussion of these technical details is not the subject matter of this book, and therefore we recommend the reader to a rule book, such as any of those written by Stallings or other acknowledged experts in this matter.

WAN Extensions

The Network Manager soon finds that the LAN should be extended through a WAN of some type. There are many ways to do this, all of them requiring some type of bridge, router, or gateway. The differences between these devices are discussed below. Suffice it to say that the epistemology involved has created several major arguments among the cognoscenti, ruining friendships, business deals, and, for all we know, marriages.

Some type of additional hardware is required to extend the LAN through use of WAN circuits. The transmission medium of choice may have some impact on what type of hardware is used for extensions. Therefore some analysis of the transmission media is required.

Choice of Transmission Media

The choice of transmission media plays a pivotal role in the preparation of the design phase. In general there are four choices of transmission media; this section will compare each of them. The four include copper, fiber, radio, and free-space light.

Copper Media

Copper is the most common transmission media and is installed worldwide. Let us examine several aspects of this choice. First it is the cheapest on a per foot basis. Twisted pair, either shielded or unshielded, is close to a commodity. The vendors are legion and, as such, the Network Manager must pay particular attention to the quality control practiced by the selected supplier. Merely stating that the product must meet a given set of specifications may not be enough. The Network Manager may wish to apply more stringent product control by requiring the supplier to provide a certificate of compliance with every reel or lot, indicating that the product meets a set of standard criteria promulgated by a committee or body which has, as one of its functions, the setting of standards for wire and cable. The Electronic Industry Association–Telecommunications Industry Association or the American National Standards Institute are two which should be considered as primary sources of standards.

Second, because of its pervasiveness, there are many, many qualified installers and not-so-qualified cable gypsies who travel throughout regions of the country doing nothing more than cable installation.

Because they know how to install telephone systems, they think that the next logical step up from telephone is twisted-pair data installation. The differences between the two technologies, however, are both subtle and obvious. Many of these firms can identify the obvious ones, but they learn the subtle ones as they go. The Network Manager may find the selected installers are doing unsupervised on-the-job training during the installation of the new network.

Third, copper media does not require specialized tools and test equipment for installation. Common hand tools, such as pliers, knives, and crimpers are all that is necessary for twisted-pair and thin coaxial cable installation. Thick coaxial cable installation is somewhat different and will be dealt with later in this section. In keeping with this discussion, copper media will allow for splices if such media are sheared or shorted after installation. The same cannot be said for other media. Copper media is also simple to install because it is neither fragile as is fiber-optic cable, nor is it technologically finicky, such as radio and light wave technology.

For very short distances—less than 100 meters—copper can carry data at speeds of 100 Mbps. For longer distances—up to 2500 meters—speeds drop to 10 Mbps or less. For WAN circuits, copper is limited to approximately 1.544 Mbps so long as there are many repeaters in the line.

Copper is not without its drawbacks. The first of these is the possibility of fatal differences in voltage potential. When coaxial cable is used, the outer shield of the cable can be at the same voltage potential as the outer case of the computer or communications equipment being used. If there are differences in wiring systems between the two ends of the coaxial network, there may be about 55 VAC existing on the coaxial cable itself. If a human gets between this and a good ground, the resulting current flow can be FATAL. When installing either thick or thin coaxial cable, it is a good idea to have a qualified electrician check out the installation for floating grounds. This is quite prevalent when the network extends across more than one building.

Second, copper can act as an antenna, picking up any kind of stray electrical field, and by doing so, inject electronic noise into the signal traversing the cable. This is particularly true when the copper cable runs near motors, copying machines, fluorescent lights, transformers, or other data or telephone cables. The fix for this problem is either rerouting the cable or putting it in a metallic conduit, then grounding the conduit. Either approach is expensive and the problems are usually not noticed until the installation is complete.

Third, copper is a metal, and, as such, is subject to attack by corrosives and rusting by the development of verdigris. In the normal office environment this is no problem. But, if the network traverses the factory floor or certain types of laboratory facilities, buildup can occur and signals cannot punch through. This is a slow, evolutionary process, one which first appears as an intermittent interruption in communications.

A final drawback is that copper is heavy. One or two runs of thin coaxial cable with several runs of twisted pair may not seem too heavy. Yet when these one or two runs increase to 10 and 20, and the several become 70, the weight starts to build and the structure which safely supported 10 is being asked to support 100 runs. Sooner or later, probably later and catastrophically, the support structure will fail.

Thick-net cable, illustrated in Figure 3-4, deserves special attention because of its physical characteristics and installation quirks.

The installers need certain additional tools to secure the connectors on the cable itself. These connectors are normally identified as F connectors and come in several pieces. The solid inner conductor of the cable must be soldered to the center conductor of the connector. The inner shield must be soldered to the outer conductor, the body of the connector. There can be no short between the inner and outer conductor. This is a difficult connection for a neophyte to make. Too much heat and the insulation breaks down; too little heat and the solder connection is not a good one and a cold solder joint results, showing intermittent connections.

Because of its physical limitations, thick net cable cannot be bent at sharp corners. If there is no room for a sweep, right-angle coaxial connectors must be used. The cable should never be coiled or curved when the circumference of the coil or arc of the bend is less than one meter. Extreme cold makes this type of cable brittle and easy to break. The inner conductor can separate without this break becoming obvious until the cable is tested for operational characteristics.

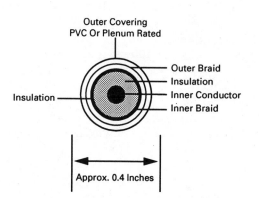

Figure 3-4. 10 base 5 cable.

One final point must be made here, even though it is not specific to copper media. The insulation used on both copper and fiber-optic cables comes in two general types: polyvinyl chloride (PVC) and Teflon™. When PVC insulation burns, it gives off a noxious and potentially life-threatening gas, chlorine. Therefore most states have laws prohibiting the use of PVC cable in plenum areas, which are the areas in suspended ceilings and in some areas under floor spaces. Teflon-coated cable is used in these areas. Yes, your fears are correct, Teflon-insulated cable is nearly four times as expensive as PVC-insulated cable. So, when doing the design and copper or fiber is the transmission media of choice, clearly identify the portion of the building(s) that are plenum-rated airspaces. PVC cable can usually be run in plenum airspaces if it is enclosed in a metallic conduit and the conduit is sealed on both ends. Which approach to use, PVC/conduit or Teflon is a pure cost trade-off. Assume the cable has to go across a hallway, a traverse of maybe four meters above a suspended ceiling. It would be cheaper to install the conduit and seal up both ends rather than run only Teflon-coated cable.

In summary, we find that copper is ubiquitous, normally easy to install, and quite applicable for most networks. It has some limitations based on distance and bandwidth. It can act as an antenna by picking up stray signals and causing disruption on the network. As it is a metal, it can corrode, causing lost data. Fiber-optic cable suffers from none of these shortcomings, yet has some of its own. What follows is a discussion of these good and bad points.

Fiber-Optic Media

Fiber-optic cable offers the Network Manager many pluses, but not without associated costs. We will examine each of these pluses and minuses in light of current and future networking requirements.

Fiber-optic cable comes in two generic types: single mode and multimode. Single-mode cable is much smaller in diameter, approximately nine microns in diameter. It is difficult to see with the unaided eye. Multimode fiber-optic cable is 62.5 microns in diameter and can be seen easily. Single-mode fiber has bandwidths in the 400 MHz range, multimode is limited to 200 MHz in bandwidth. Note that some vendors will give bandwidth for this media in bits per second in place of the megahertz metric we use. There is not a one-to-one mapping between bits and Hertz as many indicate. As a rule of thumb, multimode fiber-optic cable should be sufficient for almost any LAN functions. Single-mode fiber-optic cable can use lasers as a light source (called a launcher in the

Table 3-2. Comparison of Fiber-Optic Cable Types

Type	Size	Loss Per Kilometer	Bandwidth
Single	9 m	<2dB	400+ MHz
Multi	62.5 m	>3.5dB	150–200 MHz

terminology of the trade). Multimode fiber-optic cable normally uses light-emitting diodes (LEDs) as a light source.

Aside from the amount of bandwidth delivered, the Network Manager must realize the single greatest difference between single mode and multimode cable is the loss on a per meter basis. Table 3-2 compares the two types of fiber-optic cable on several criteria.

As an application note, the Network Manager may be interested to note that most single-mode fiber-optic cable in use today is that used by telephone companies, both LEC and IXC for trunking between switches. The lower power loss of the single-mode cable makes budget planning for power much more attractive. Interestingly enough, both types of cable show about the same amount of loss at a mechanical connection. The controlling factors appear to be the quality of the connector and the skill of the technician making the connection.

A very large plus on the side of fiber-optic cable is that it is immune from outside electrical interference. Fiber-optic cable can run next to power transformers, x-ray machines, or other high voltage-high current sources without injecting stray signals into the traffic flowing over the transmission media. Unlike copper, fiber does not conduct electricity. If two ends of the cable are at different electrical potentials, no harm can come to human life. A minor, but still useful point of fiber is its weight. On a purely bandwidth per pound basis, there is no comparison, assuming that two strands of multimode fiber-optic cable run in one common sheath. Weight on a per meter basis is somewhere in the single gram range. If these two strands (one transmit, one receive) can carry 150 MHz worth of traffic, 15 runs of copper would be required, each in the tens of grams range for the equivalent bandwidth.

Fiber, however, is not without its drawbacks. First, it is fragile. We have seen cable installers hook copper cable to the rear bumper of a truck to pull it through an underground conduit. Fiber would break under this type of stress. Some of the more careful installers have a device which starts to slip when too much stress is put on the fiber cable itself: The concept is similar to a torque wrench. Secondly, fiber cable cannot take very sharp, short bends. The individual strands within the cable itself will crack, or break, and destroy conductivity.

Cracks or bends become much more noticeable with single-mode fiber-optic cable than with multimode.

Another shortcoming of fiber-optic cable is that the connectorization of this media is not for the technical neophyte. Certain specialized tools are required. The actual connectorization is a finicky process; one microscopic speck of dust will render a connector useless, and the process will have to begin all over again. With the cost of the connector almost equivalent to the cost of one hour's labor for the installer, the Network Manager soon sees the problems the construction crew will run into. Microscopes and specialized test equipment are needed in the final test of the connector itself. Consider the problems if the two ends of the cable are hundreds of meters apart. How do the testers coordinate their actions? Once the fiber is installed, the actual mechanical connectors must be kept enclosed in dust-proof containers. Dust or dirt specks getting in between connectors will act as a block, preventing light from flowing through the strand.

There are two types of splicing of fiber-optic cables: mechanical and fusion splicing. Mechanical splicing is the faster, cheaper, and less labor-intensive approach. Mechanical splicing will create a loss of somewhere between 0.6 to 1.3 dB per pair of connectors. A fusion splice is much more expensive and labor intensive. A fusion splice actually melts the two conductors together at heat high enough to melt glass. In counterpoint, a good fusion splice will create a loss somewhere in the 0.01 to 0.3 dB range. Many of these variances in physical connector loss is in the skill of the technician.

Even with the lower limit of bandwidth offered by multimode cable, the Network Manager can feel reasonably certain that a fiber-optic link or network will carry all the traffic generated by the computers attached. Consider: One Ethernet normally carries traffic in the 6 to 7 Mbps range. It would take twenty of these networks to fully load one pair of fiber-optic strands.

In summary, with the exception of the cost of installation and physical fragility of the fiber-optic strands themselves, fiber seems to be the ultimate transmission media. But, like any other physical device, there are some places it cannot go. The next sections look at two other transmission media and what role they play in the design phase of LAN procurement.

Radio as a Transmission Medium

Radio communications in LANs fall into two categories: point-to-point links, normally running at microwave frequencies, and broadcast, running in the 800 to 900 MHz range. We will examine each of these two with an eye to both good and bad points of each.

Microwave communications normally falls in the 2 or 6 gigahertz (GHz) range. Some microwave transmission also takes place in the 18 and 23 GHz range. As a rule of thumb, the higher the frequency, the shorter the 'hop,' or the distance between a transmitter and receiver. The ultimate distance the user can plan for is roughly 30 miles, 50 kilometers, or line of sight.

Planning the physical siting of microwave facilities is not for the unskilled. The Network Manager is strongly recommended to select a competent consulting firm for this. There are two major problems in such planning—one technical, the other legal. The technical problem is determining if there is a path between the transmitter and the receiver. Even if no mountains are in the way, a building may interfere with the transmission path. An obstruction such as a building or hill interrupts the line of sight necessary. As a side issue, there may be frequency interference between existing microwave facilities and planned additions. The federal government, responding to international legal requirements, is eliminating parts of the 2 GHz spectrum for commercial usage. Legal problems include such things building ownership or zoning laws. If the Network Manager's firm is leasing the premises, the landlord may not allow installation of an antenna on the roof or side of the building, and zoning laws may prohibit the installation of a tower upon which the antenna is to be mounted. One further note of caution: Microwave energy can be life threatening. No one should ever walk in front of the antenna when the transmitter is turned on and operational. The concept is the same as a microwave oven.

The plus side of microwave usage is that once the initial costs of procurement and installation are amortized, the transmission costs are free. Many vendors offer microwave radio systems, capable of fully digital transmission at 10 Mbps, which is a full Ethernet network speed. Modern microwave systems are small and unobtrusive, no larger than a two-drawer filing cabinet. Many of them require no special equipment rooms and have a mean time between failure in the five year range.

Radio communications in the 800 to 900 MHz range are much more limited in nature and are broken down into two separate categories based on licensing procedures. These frequencies in the sub-gigahertz range include that band which requires a Federal Communications Commission (FCC) license and that which is unlicensed. This difference is brought up for the Network Manager's consideration and then dismissed. Most vendors of equipment that requires licensing will provide the license at a minimal or nonexistent fee.

The key point to remember is that such technology is really local. A range of 500 feet is stretching the envelope. These devices are designed to work in a bull pen or cubicle arrangement where the room or offices are separated by no more than 350 to 450 feet. The bandwidth of most of these devices is limited to low kbit per second range, so they are not useful for large file transfer or graphics applications. The lack of bandwidth and distance are not the only problems associated with sub-gigahertz radio. The first is the problem' with interference. Any human-caused or natural atmospheric disturbances will drastically slow transmission or even eliminate it altogether. Therefore external communications paths, or paths which cross a factory floor, are not good candidates for this technology. A path of propagation that happens to go over a microwave oven, well, such problem that is the "meat and potatoes" of engineering firm or consultant. Second, because the signal strength is so weak, a fire break of brick, concrete, or even heavy gauge metal will prohibit others from receiving the signals. The Network Manager should also consider the fact that these signals are broadcast. Anyone with a sensitive receiver could intercept the signals and look at information such as spreadsheets or internal corporate correspondence.

The only good point to a sub-gigahertz radio is the fact that no wires are involved. If subscribers move frequently, the Network Manager does not have to move the connections with them. When the computer moves, the link moves with it because the hardware is nothing more than a PC board with an attached antenna.

In summary then, we have two types of radio communications: Point to point and broadcast. Point-to-point microwave communications provides good backbone linkages but is not for the beginner. It is cost effective and reliable as an alternative to leased lines. Sub-gigahertz communications is simple but limited in range and throughput. This leaves us with the last transmission media, light.

Free Space Light

The technical specifications of free space light as transmission media are quite similar to those of microwave and sub-gigahertz transmission. Point-to-point and broadcast transmission are the two modes of operation. We will examine the point-to-point mode first.

This evaluation begins with ensuring that a line of sight exists between the transmitter and receiver. No special expertise is needed here as the Network Manager must be able to physically see the other end. Distances between the transmitter and receiver should not exceed a manufacturer's specifications. We do not recommend spans of over 1 mile or 1.2 kilometers. The transmitters and receivers must be solidly locked in place. Deviation of an inch or more will create a loss of signal and its concomitant interruption of data flow. Heavy fog, rain, snow, sandstorms, dust, or other types of atmospheric pollution also will interrupt communications. Basically, the only place where such technology would be usable is within a very small campus arrangement, or from building to building within an urban area. The equipment is small, light, and normally quite trouble free. Bandwidth is in the full 10 Mbps range and requires minimal tweaking once the network is established.

Broadcast mode is almost exactly parallel to sub-gigahertz radio propagation. Bandwidth and distance are limited, bandwidth in the tens of kilobits per second range and distance is under 400 feet. One of the problems with this approach is that any solid, opaque substance will eliminate the signal. A paper bag over the emitter or receiver will cut transmission totally. In counterpoint, this type of transmission media is almost designed for the shop or factory floor because the only thing that can generate interference is something which appears between the emitter and receiver. The technology of transmission is the same as that used in multimode fiber-optic cable, so it is well understood.

Review

This will be a good time to review our discussion of transmission media as such things are a major part of the design phase. Copper comes in two general types: coaxial cable and twisted pair cable. Both types carry the same amount (throughput) of data; it is the distance to be traversed that is the deciding factor. If more than 100 meters of run is needed, the Network Manager should be looking to coaxial cable, not twisted pair. Fiber-optic cable provides the most bandwidth but is also the most technically challenging to install and maintain. Radio provides connectivity where other methods are not appropriate or financially acceptable. Free space light is quite analogous to radio as a transmission media, but is of much less range. Each of these four media has strengths and weaknesses in the overall design, and the Network Manager must mix and match these to fit the organization's needs.

Connectivity

Once the Network Manager has decided on a transmission media, then the next decision is the connectivity of the network. The connectivity will be self-determining: Form follows function. What follows is a discussion of four areas the Network Manager must consider when selecting the network connectivity. Note that this discussion does not include proprietary technology, such as IBM's Systems Network Architecture (SNA), Digital's DECnet, or Proteon's ProNET. This exclusion is based on the fact the vendors of each of these topologies will be more than glad to tell you about them. In detail—great, exhaustive, b-o-o-r-ring detail. We will discuss the pros and cons of non—proprietary systems in general, the International Standards Organization's OSI model, offerings from the Institute of Electrical and Electronic Engineers (IEEE), and finally a discussion of Network Operating Systems.

Proprietary versus Nonproprietary

Proprietary connectivity systems in general are acceptable. If they were not, they would not survive in the marketplace. We have no axe to grind with IBM, Digital (DEC), or Proteon. All are fine firms with marketable products and services. But the Network Manager must understand that IBM's functional approach to connectivity may not be similar to DEC's. This proprietary approach prohibits the free and uninhibited flow of data, information interchange. The Network Manager must consider the need to reduce data and connectivity to the lowest common denominator if such proprietary systems require interconnection; an unpalatable approach to a new installation.

Proprietary systems also have the problem of support. One of the pioneers in office automation and an early user of LANs, was the WANG corporation based in Lowell, Massachusetts. WANG's proprietary connectivity, WANGnet, would not "talk" to anything except their product line. Owners of WANG equipment could find they have "orphans" in terms of growth and support. None for the former, and increasingly expensive for the latter. While we do not imply proprietary systems were the cause of WANG's downfall, even major vendors find that supporting obsolete or outdated connectivity systems are not high on their list of activities.

Based on this lack of openly available support, the Network Manager must consider that the support of proprietary systems is a problem which could occur if not in the immediate future, within ten years of installation. No one wants any more problems than is absolutely necessary. The opposite of proprietary systems are those offered by vendors complying with the OSI model.

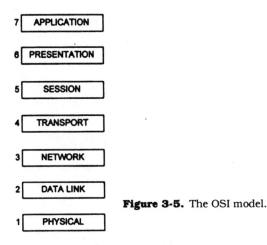

Figure 3-5. The OSI model.

Open Systems Interconnect

The OSI model is diametrically opposite to the proprietary approach. In the OSI model, any vendor can build a product to meet certain publicly documented specifications. When the product is certified by independent laboratories to meet OSI specifications, the Network Manager can assume it will work with every other OSI model. The OSI model is in seven layers, or sets of specifications. Shown above is the commonly accepted diagrammatic way of looking at the OSI model. We discuss each layer very briefly following the order of Figure 3-5.

Layer 1. Physical Layer. The physical layer is responsible for the transmission and reception of bits over the transmission media. Data entering the physical layer is normally in a serial stream of bits, not in parallel streams as is present in the computing device. The physical level identifies the electromagnetic connectivity between the computer and the network. It also describes voltage and current levels or modulation techniques.

Layer 2. Data Link Layer. The data link layer creates packets of data of the appropriate size to transmit over the network. The actual size of the packet is a function of the parameters agreed upon between the sender and the recipient before the actual data flow takes place. This agreement can take place in other levels of the ISO model except the physical layer. Much of the error checking of data integrity can take place at this layer. The process of packetization is reversed when data arrives at the intended recipient. Some authorities claim error detection and correction through retransmission takes place here.

Layer 3. Network Layer. The network layer ensures end-to-end delivery within a single network. This is done through a technique called addressing, where the intended recipient—the sink—and the sender—the source—have unique alphanumerical addresses. The entire concept and application of addresses is so detailed not a few doctoral dissertations have covered it to one degree or another. (Addresses also play an important role in the transport layer.) Congestion control is another part of the network layer.

Layer 4. Transport Layer. As the network layer is responsible for end-to-end delivery within a single network, the transport layer is responsible for delivery across multiple networks. Some writers claim the network layer does not do sink to source destination routing. We disagree. In a single network, the domain of the network layer, sink-to-source routing does take place. The transport layer subdivides large packets from upper levels for use by lower (network and data link) layers.

Layer 5. Session Layer. The session layer is responsible for initiating, controlling, and terminating a communications session between two OSI-compliant sinks and sources of data. Some writers claim name-to-address conversion takes place at the session layer. We do not disagree but find that name-to-address conversion normally takes place in bridges or routers, a layer 3 or layer 4 device, respectively.

Layer 6. Presentation Layer. Layer 6 provides data translation, for example, between one spreadsheet format and another. Although encryption is not a normal concern for the Network Manager, it should be noted the encryption and decryption process normally are thought of as taking place at layer six. Some of the more recent writers indicate the "White Pages" (X.500 directory service) will take place at layer six.

Layer 7. Application Layer. The application layer is just what its name specifies; applications that people use take place here. One of the more common is the virtual terminal emulation protocol, VTAM.

As a guide to understanding the OSI model, the Network Manager may want to consider layers 1 through 4 as the domain of the technician and low-level engineer. Layers 5 through 7 are the responsibility of the programmer or experienced software engineer.

The Network Manager also may want to consider that the OSI model is just that, a model. Various vendors claim that their product supports the model or that their product is OSI compliant. Compliance is to a certain extent like the difference between alive and dead. One can be partially compliant with the OSI model, but not partially dead.

The Network Manager must take great pains to ensure that the product is fully compliant with the most recent specifications of the OSI model. Now we will review a more fully developed approach to specifying connectivity, the IEEE series of protocols.

IEEE Protocols

There are three major IEEE protocols the Network Manager may want to evaluate for applicability. These are Collision Sensing Multiple Access Bus (IEEE 802.3), Token Bus (IEEE 802.4), and Token Ring (IEEE 802.5). Of the three, we do not remember seeing a lot of hardware and software that supports IEEE 802.4 LANs. That is not to say such LANs do not exist. ARCNET is a token bus, that supports traffic at 2 or 20 Mbps, but some experts claim a separation between ARCNET, an early product of the Datapoint Corporation, and IEEE 802.4. We do not downplay the applicability of either ARCNET or IEEE 802.4 to a network. ARCNET is a venerable technology and is quite capable of passing bits at a high rate for long periods of time. However, for the purposes of this book, we will concentrate on those technologies that have the major market share, IEEE 802.3 and 802.5. A few notes of caution.

Xerox's Ethernet (a trade name) and IEEE 802.3 are *not* interoperable. We offer in support of this position "Ethernet Vs. IEEE 802.3: There Are Differences," by Gary Kessler in the July 1987 issue of *LAN* magazine, pages 54 to 57. Mr. Kessler cites some sources in the article if further research is desired. In counterpoint, the IEEE 802.5 standard is very close to IBM's proprietary token ring product. We have never found anyone who can point out significant operating differences between the two. With these observations behind us, let us continue to consider the choice between IEEE 802.3 and IEEE 802.5 standards as the selection for the new network.

We will begin with the IEEE 802.3 topology. It is a bus network, with free-for-all (heuristic) access. A station will "listen" to the network, and if it senses no traffic, it will transmit a packet. This continues until it has completed sending all packets or until it detects another station transmitting in error. Both stations then stop, wait for a random period of time, then retransmit again. IEEE 802.3 networks are best used when many small (less than 100 kbits) transmissions are required. Subsets of IEEE 802.3 are popularly known as 10BaseXX where the XX is the separate designator. These different "flavors" are discussed below.

10Base2. 10Base2 is sometimes known as ThinNet or CheaperNet. It uses RG58 coaxial cable, which is about 0.25 inches in diameter. The connectors are twist on—BNC—with crimped connection to the wire. This flavor is limited to about 165 meters in length and no more than 30 stations attached. Stations must be separated by 0.5 meters. We have heard two "experts" argue the throughput is either 2 Mbps or 10 Mbps. If it is 10 Mbps, why is it called 10Base2?

10Base5. 10Base5 is the more traditional Ethernet installation which uses thick coaxial cable that is roughly 0.5 inches in diameter. Connectors are F type, requiring skilled technicians to install. This flavor is limited to 2500 meters in length, with a total of 500 (roughly) stations attached. Stations must be separated by 2.5 meters. It seems that there is a contradiction here: If we divide the maximum length of the network (2500 meters) by the minimum separation (2.5 meters) we come up with a possible 1000 stations on the network. Actually, the theoretical upper limit of an IEEE 802.3 network is 1024 stations. The figure of 500 stations is one quoted by several knowledgeable authors and is a good working figure. The maximum throughput is rated at 10 Mbps.

10BaseT. 10BaseT is the unshielded twisted-pair wiring format used to run 10 Mbps over relatively inexpensive, twisted-pair copper wire. The length limit for 10BaseT is 100 meters, with only two stations on the network. The connector is similar to the RJ45 connector used by telephone installers. It meets certain higher specifications in regard to pair-to-pair crosstalk. This arrangement is normally a user to hub, which is more of a star than a bus topology. Throughput is rated at 10 Mbps. A subset of this flavor, a fiber-optic interface repeater link (FOIRL), increases transmission lengths to roughly 2 kilometers. FOIRL is becoming more prevalent with the widespread use of 10BaseT.

10Broad36. The 10Broad36 is a seldom seen IEEE 802.3 flavor as it runs the data over the same system which is used to carry cable television (CATV) signals. Some experts limit the length of the network to 3600 meters. The actual distance will vary depending upon the hardware vendor. The number of stations is dependent upon the transmission media, not the format of the signals. Throughput is rated at 10 Mbps.

IEEE 802.5 has two flavors, the 4 and 16 Mbps rates. Many of the same rules apply to both flavors. A single network is limited to 72 stations. The stations may not be further from the media access unit (MAU) than 120 meters. Using repeaters, the network can be extended, but each repeater suffers the 120-meter limitation. Most organizations get around this by connecting MAUs in a separate ring. Two MAUs can be separated by roughly 230 meters. It is important to note that this length restriction can be overcome by using fiber-optic cable as a

Table 3-3. Topology Comparisons

Characteristic/ Network type	10Base2	10Base5	10BaseT	IEEE 802.5	IEEE 802.5
Throughput (Mbps)	2	10	10	4	16
Length using copper (meters)	138	500	100	120	120
Length using fiber (meters)	2500*	2500	N/A	2000	N/A
Maximum stations (per segment)	30	200	N/A	72	72
Total maximum stations	30	1024	N/A	264	264

*Fiber or copper

transmission media. In this instance, the maximum circumference of the ring is 2000 meters. See page 54 for the selection of fiber-optic cable as transmission media.

The IEEE 802.5 topology allows a station to transmit by passing usage authority in the form of a token to the next station in line. See Table 3-3. The station holding the token (a special pattern of bits) releases it after transmission. The token circulates to the next station and the process repeats. While this process allows greater throughput under heavy loads, it is, as the Network Manager may well intuit, wasteful under light loads.

Table 3-3 shows a comparison of these two technologies.

With all of these technological facts in mind, how does the Network Manager select the appropriate standard for the new network? One possibility is to throw out the technical aspects all together and consider these questions.

1. Which technology is more expensive, IEEE 802.3 or IEEE 802.5? Do not fall into the trap of bits per second versus dollars—there is no comparison. Just because the Network Manager pays more does not mean the hardware is better. Consider also the software used to drive the hardware. Some network operating systems (the subject of the next section) do not work over one topology or the other.

2. What is the availability of new or additional hardware? The larger the installed base in the marketplace, the better the chance of new vendors coming on line with improvements or others retaining their market share.

3. What about test equipment? Once the network is installed, someone must test it. If no test equipment is available for this hardware, how can that be done?

4. How does the topology fit into the network management picture? Formal network management is discussed on page 36.

Review

To summarize, there is very little rational reason to place one standard above the other. As of this writing, several groups have formed to offer new products in the 10BaseVQ or 10Base100 flavor of IEEE 802.3, offering 100 Mbps over unshielded twisted pair. The answer to the 802.3 versus 802.5 question is....

Network Operating System

Once the Network Manager has decided which topology to use, the next decision is that of which network operating system (NOS) to select. Currently this software, and in some cases hardware, comes in two configurations: server based and peer based. Comparing the advantages and disadvantages of the two is comparing apples and oranges, or more closely, the family sedan and a high-performance road course machine. Both will get from point A to point B, but what a difference in the ride!

Let us step back from the nitty-gritty of the decision process and review what we are trying to do with a network. As is stated in the introduction of this book, a network is the means by which information interchange takes place. Bits and bytes are important, but not as important as transparency to the user. Whereas we enjoy technical delving and tweaking, the users are more concerned with getting their work done; memorizing esoteric command line functions is seen as a chore, not a joy.

Consider the results of the following steps concerning the connection of a new user to the server. (Assume the network to be in existence.) The systems administrator must

1. Create a user account

2. Assign password(s)

3. Set privileges regarding file and directory creation

4. Demonstrate log-on procedures

Does this sound like the steps necessary to put a new subscriber on the server in the organization of the 1990s? That is unusual because that is the same procedure used when we were first introduced to computers in the early 1970s. If memory serves, it was a PDP-11(8?).

What then are the benefits of a client-server model of computing? Essentially we see very few; it is a lot like cancer, it just grew. The idea of taking information out of the domain of MIS, the glass house, and to the user's desk seems to have stopped along the way. Instead of a visually impressive piece of hardware with multiple blinking lights we have a bland box, usually set in a dark room, quietly whirring away storing our efforts and hopes, our checking accounts and secret letters, and even some work-related information. If this is the case, and it usually is, what are the parameters for choosing the box?

The NOS should support DOS, Windows, MacIntosh, OS2, and perhaps UNIX. While not all NOSs support all these operating systems, those which do are high on the priority list. Other operating system parameters include the following.

1. *User administration.* Name and directory services are a minimum for mail/file forwarding. Authentication can be on a per server or netwide basis on initial log on. A service should appear only once, regardless of which server it resides upon. Servers should be accessible by name, not some abstract network address.

2. *Server administration.* Each server can have an administrator or superuser identified for maintenance and operations. The administrator can create and manipulate groups of users. For example, all users in Group XXXX can access printer number 14 only, the one located in their field office. The server administration function enables directories and files to have varying levels of security: No more than four levels are necessary. Servers should be capable of administration from remote locations. There should be no need for the administrator to have to travel around the country or even around the block. Backup capabilities should be inherent in the server itself.

3. *Telecommunications protocols.* The server should, with the appropriate hardware and software, operate in all telecommunications protocols, holding multiple stacks in memory for each necessary protocol. The server also should provide routing support. Interfaces to common network management protocols also should be part of the server software suite.

4. *Third-party software.* Many network operating systems are not, in and of themselves, sufficient. Other firms make money by selling third-party software which remedies these shortcomings. The server must support these additional pieces of software without memory conflicts.

These four areas only scratch the surface of the decisions the Network Manager faces when designing the LAN. At this point in the process it is reasonably certain what operating systems will be in use—DOS, UNIX, or something else. The network operating systems—Novell's *Netware,* Microsoft's *NT,* or Banyan's *Vines*—are three of the more common ones. What applications are going to run? Personal productivity software, such as word processing, spreadsheets, databases, and graphics must reside somewhere. Is it better that the application software resides on the server or in the user's workstation? We will discuss some of the points concerning this question.

By keeping the applications software on the server(s) the Network Manager reduces the effort necessary in updating the applications and ensuring that all copies are concurrent. On the other hand, the Network Manager is putting all the eggs in one basket: If the server fails, then no one has access to the applications necessary to do business. On the other hand, by keeping applications at the subscriber's machine, the Network Manager ensures that failure of the server means failure of telecommunications, not failure of information interchange. Floppies and tape can still pass data from one person to another. In counterpoint, the Network Manager must worry about applications being all of the last (or at least of some common) revision. Pirated software and uncontrolled proliferation of unpaid-for software opens the firm up to lawsuits for operating software without a license.

User files kept on the server will soon clog even the largest hard disk, slowing data storage and retrieval. Further, the Network Manager must create and implement some type of policy concerning the backup and removal of files over a certain age as well as assigning storage space. A crashed hard disk, after all, will be blamed on the Network Manager, leaving the subscribers irate and business in the lurch. In counterpoint, keeping files on the subscriber's machine may be a better approach, but how sensitive is the information being kept in an unsecured machine? Making the users responsible for backing up their files is a good idea, but there may be some disparity between the theory and application of this idea.

So, where is the place the Network Manager must stand to make this design decision? Somewhere between the two extremes discussed above. The Network Manager must know the organization, its needs, desires, and corporate culture. A consultant can offer a *technically* correct decision, but not one that is applicable to a given organization. In this case there is no textbook answer to the design approach used by the Network Manager.

The client-server approach discussed above does not provide true distributed computing, but rather a new, more technically sophisticated update of the terminal to mainframe paradigm. The systems administrator no longer is concerned with large mainframes but smaller boxes scattered throughout the organization. The selection of software is no longer limited to the hardware vendor but from multiple "here today and gone tomorrow" software houses. The power in the subscriber's hands is enough, if improperly used, to bring the server and the network crashing down. True distributed computing is shown in the peer-to-peer networks, which are discussed next.

In a peer-to-peer network, the Network Manager still has a place and a mission, but it is more supportive than managerial in nature. The Network Manager has no single machine to worry about but many small- to medium-size machines scattered throughout the enterprise. There are several areas of concern when selecting the peer-to-peer approach, and they are discussed below.

1. *Operating system support.* The NOS should support DOS, Windows, a nonmemory intensive version of OS2, and perhaps UNIX. Not all peer systems support all these operating systems. Still, the operating systems in use are high on the priority listing.

2. *User administration.* Unlike the client-server paradigm, there is no need for user administration or server administration in a peer network. Each person "administers" his or her own machine, being responsible for file backup, file security, and privileges. The Network Manager must consider software distribution to be the only major problem with a peer system.

3. *Telecommunications protocols.* The server should, with the appropriate hardware and software, operate in all telecommunications protocols; such peer systems, however, are the exception and not the rule. Most peer systems support only IEEE 802.3 or 802.5 as telecommunications protocols. Many peer systems have their own, proprietary protocols and as such are suspect for a new network.

4. *Peer servers.* Some peer systems offer support to a file server as a large-scale storage device. Many of the same problems and questions concerning the client-server model arise based on this approach. In a true peer system, each subscriber performs information interchange as they need or are requested to do by their fellow employees.

Unfortunately peer systems are not as well thought of as are client-server systems. Of course this parallels the acceptance of network operating systems in general. With that observation, let us summarize the comparison between the two types of network operating systems.

The client-server model is thought well of and has a very large share of the market. It will be with us for years to come. It does the job required, although not always gracefully, and with user friendliness. Peer systems are much more user friendly, but nowhere near as powerful as the client-server systems. In counterpoint, the peer system is easier to operate once it is installed. The client-server approach requires constant tweaking and grooming, particularly in the storage of software applications and their files. Client-server offers more security and telecommunications support but at an increased dollar-per-machine cost.

The Network Manager also should be aware of the "religion" problem concerning this decision. Many subscribers were brought up on certain operating systems, network operating systems, and applications. An old hand at UNIX will not want anything but TCP/IP and VI for a text editor. If such persons do not like a particular application, they will write their own. On the other hand, a PC-DOS-oriented person will not even think of writing a simple batch file in EDLIN when they can spend $500 to buy a new killer application just to rename old Lotus 1-2-3 files. If it runs with some graphic user interface (GUI), well, so much the better, and if it takes up 8 Mb, who cares? Ah yes, then there are the MacIntosh users. Hmm, well, if necessary we can do something for them as well. SNA and VMS gurus are beyond the pale and can be safely ignored with the rest of the Luddites. Enough of this wallowing in the morass of opinions. It is time to forge onward, or at least stumble slowly, to an area which purports to have rules—the selection of telecommunications hardware to support the design. Or, phrased differently, how I learned to live with bridges, routers, and gateways.

Network Extension Hardware

Different types of devices can be used to extend LANs beyond the very short distances defined by their respective topologies. Much confusion exists concerning the differences between these types of devices and where they belong in the LAN topology. What follows presents a possible descriptive solution to the problem.

Considerations

For this section certain shorthand techniques are used. The IEEE 802.3 label will be used for all CSMA/CA or CSMA/CD topologies. The IEEE 802.5 label will be used for all token ring–based topologies.

Four questions must be answered by the Network Manager when selecting the appropriate device for connecting multiple networks of the same or different types.

1. *Networks installed.* Are networks of like types being connected to each other, or is this an inter-network, with networks of unlike types? The term *types* in this instance indicates network topology.

2. *Protocols in use.* Do all networks use the same protocol: TCP/IP, X.25, or other universal or proprietary protocols?

3. *Operating systems.* Does the data terminating equipment (DTE) at both ends use the same operating system: DOS, UNIX, Novell's *Netware*, or Banyan's *Vines* for example? One assumption being made here is that the network is used for more sophisticated applications than mere file transfer.

4. *Network size.* What is the actual length of the network? Remember, these networks are constrained by their topology. IEEE 802.3 networks are limited to 500 meters per segment; maximum "book value" length of 2500 meters with repeaters. We have heard of IEEE 802.3 networks extended to 2800 meters, but it is not a recommended practice. IEEE 802.5 networks are constrained to a distance of 120 meters without repeaters. Using repeaters and fiber-optic cable these networks can be extended to 2000 meters.

Decision Matrix

As noted in Table 3-4, the selection of which type of device to use is almost as simple as reading down the left-hand column, then across to the appropriate device. For example, if the network type, protocol, and operating system used were all the same, yet the network size was greater than the maximum physical length allowed, then remote bridges would be used to link the two networks.

The reader also should be aware of certain recent changes. Other authors have, in the past, used the type of transmission media (e.g., thick coax, STP/UTP, fiber, or thin coax) as a criterion for defining dif-

Table 3-4. Decision Matrix for Hardware Selection

Device Type	Repeater	Bridge	Router	Gateway
Parameter				
Network type	Same	Same	Same or different	Different
Protocol used	Same	Same	Same or different	Different
OS used	*Same*	Same	Same	Different
Physical size	<Maximum	>Maximum	NA	NA

ferences among these four types of devices. The arrival of network hubs from various vendors has eliminated these descriptive criteria. Wiring hubs translate from one transmission medium to another without outside assistance. Rigorous application of the definitions discussed in this article will go far in identifying whether such wiring hubs are to be repeaters, bridges, routers, or gateways.

Unfortunately some vendors have clouded this issue by calling their products "brouters," "bridging routers" or other confusing names. A remote bridge links two networks of like types, protocols, and possibly operating systems when the total distance is beyond what is allowed by the rules of network topology. See Figure 3-6. The key point to

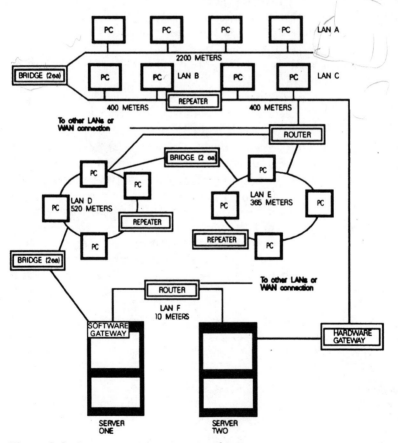

Figure 3-6. A perspective on internetworking.

remember is bridges *cannot* control the direction of the packets passing through them. Bridges, like repeaters, must retransmit every packet they receive. Routers on the other hand do not have this constraint.

Routers look at the destination address of every packet that comes to them and decide which packets belong on which network. Packets with a destination on the local side of the router do not enter the remote side network. Packets with a remote address are removed from the local side of the network so they do not increase traffic loading. So far, so good. Now let us look at gateways.

Many people think of gateways as very expensive and esoteric devices; probably a perception that can attributed to the vendors. Trick question: What is the most common type of gateway in existence? (Hint, you can mail order them for about $59 in most computer supply catalogs.) Give up? A modem. A modem can be used for file transfer between different networks, both using different protocols and different operating systems.

As we are concerned with higher levels of functionality than mere file transfer, let us look to another possible device. The X.25 packet assembler/disassembler (PAD) also can be described theoretically as a gateway, although it cannot be thought of as a router. Please note that a router is a device that identifies packets for forwarding; a PAD forwards everything it gets from either side of the network. Yet, as we wish more than the simple file transfer paradigm, we must discard the PAD from consideration as a gateway. Most PADs do not perform the operating system translation necessary for proper gateway functionality.

So we have defined what is *not* a gateway. What then is a gateway? Modems and PADs cannot be truly described as a gateway. A gateway does one thing no other device does, it translates operating system commands. Gateways can be created in software. Look again at Figure 3-6. The file server number one is running UNIX and is acting as a storage device for multiple DOS-based files. The commands that affect this file server come in through TCP/IP from DOS-based clients. The file server translates these commands to UNIX commands in order to modify files within itself. It then translates the resulting changes back to a format understandable to DOS-based clients before placing them on the network. This is accomplished through a vendor-supplied software package. All the DOS-based client user sees is a large assembly of virtual hard disk drives. The same functionality can be completed through hardware and firmware.

In summary then there are a few very rough and admittedly incomplete rules of thumb to apply to device nomenclature. Does the device

- Forward packet within a single network? It is a repeater.
- Forward packets between like networks? It is a bridge.
- Evaluate address for local or remote networks and forwards packets? It is a router.
- Perform operating system translation? It is a gateway.

MODEMs and PADs should be thought of as bridges.

Inter-networking

Referring again to Figure 3-6, let us see why some devices are used where they are shown, and why others are not. Please pay careful attention to the physical sizes of the networks as they are important.

Although not clearly shown in Figure 3-6, every device labeled "Router" may be one or two boxes, one at each LAN depending upon separation and other physical details which will vary from LAN to LAN.

LANs A, B, and C. These are three IEEE 802.3-based LANs. LAN A has five repeaters in it, which are not shown in this diagram. Remember, an IEEE 802.3 LAN is limited to 500 meters per segment. Repeaters are needed to connect these segments. This is shown more clearly where LANs B and C are connected through a repeater. Another key point to remember is that a single IEEE 802.3 LAN is limited to 2500 meters in length. The use of the bridge between LANs A and B overcomes this limitation.

LANs D and E. These are two IEEE 802.5-based LANs. LAN D is 520 meters in circumference. This is beyond the 120 meter limitation of the IEEE 802.5 specification. Therefore a repeater is necessary to extend to this limit. Like the use of the bridge between LANs A and B, a bridge is required between LANs D and E. Why? First, because of the distance involved. These two LANs could be one floor apart in a single building or be on both coasts of the United States. The bridge is necessary to overcome the timing problems (among others) with IEEE 802.5 networks.

LAN F. This approximately 10 meter LAN can provide redundancy between networks. Note that in Figure 3-6 LAN C is connected to Server 2 through the hardware gateway. Assume that gateway fails. Connectivity then flows from LAN C to LANs E and D to Server One, then through Server One to Server Two, its intended destination.

Servers One and Two can pass traffic to each other through the router. It will pass traffic with their respective addresses in both directions. Still, if the hardware gateway fails, the router, and the router between

LANs C and E can be reprogrammed to pass all traffic in both directions. If properly done, this reprogramming should take less than 60 seconds, depending on the operating system in use and prior planning.

Failure Mode

Network Managers should use caution when using this approach at redundancy for two reasons. First, if either gateway fails, the traffic flowing through the IEEE 802.5 LANs will increase tremendously. The reader will note that the router between LAN C and LAN E acts to filter all traffic for Server 2 away from LAN E.

Second, it should be realized that all repeaters, bridges, routers, and gateways are single items, and this means these are all single points of failure. When, not if, they fail, restitution must proceed through rerouting or replacement. Where possible, hot spares should be colocated to speed restitution.

Review

In summary then, this section has given a brief description of repeaters, bridges, routers, and gateways. It has attempted to describe their place in the hierarchy of inter-networking. We must point out that many intelligent, well educated and experienced people in the field of networking will vehemently argue that a bridge can filter, and thereby route traffic. There is a difference between discarding a packet, an action taken by a filtering bridge, and routing a packet, an action taken by a router. We offer this simple method of telling the difference between the two devices. If it has one input and one output port in use at the same time, it is a bridge. If it has two or more input or output ports in use at the same time, it is a router.

At this point, the Network Manager should be able to make decisions in the following categories:

- The amount of traffic flowing between elements of the organization
- The physical distances between elements of the organization
- The appropriate transmission media to use between subscribers
- The topology of the network
- The network extension hardware that may be necessary

From this information the Network Manager creates the test plan for the new installation. What follows is a discussion of the test plan and what it actually measures.

Test Plan

The test plan is not definitive by nature. The testing procedures are application dependent, and therefore can only be written in the RFQ. Evaluation of the attached RFQ will provide definitive details for that particular proposal. What follows is a high-level, almost generic approach to a test plan.

Testing plans are divisible into four separate generic parts or layers following the OSI model. Testing will take place at the physical layer, the network layer, the transport layer, and the application layer. What follows is a very brief discussion of this layered approach.

Testing at the physical layer includes tests on the transmission media, connectors, and power sources. Testing at the network layer applies to telecommunications within one network identified by a subnet mask in the TCP/IP protocol suite or by one block of IP addresses (NNN.NNN.NNN.000-255). Testing at the transport layer concerned with LAN to LAN connectivity, either through routers alone or through routers and the WAN circuits linking them. Application layer testing is concerned with end-to-end connectivity and operation between end systems across the LANs and WAN. This ensures there is no conflict between applications software, sinks and sources of data, and the network(s) connecting all these.

Some or all of the following tools and test equipment are necessary to complete tests. The specific details for each of these are mentioned in the discussion of each testing procedure which follows. It is important to note that these tests do not troubleshoot for a defective protocol stack or bad connector, the tools only show that such a condition may exist. Tools are broken down into two groups, those that a Network Manager must have and those it would be nice to have.

Must Have

- A device for checking shorts, opens and pair transpositions. A digital multimeter available at most electronic retailers for $29.95 or less is the minimum sufficient.

- A stopwatch accurate to 0.1 second.

- A laptop personal computer with terminal emulation software and protocol stacks for every protocol used in the network(s) to be tested.

- Test files, normally ASCII, which are 10 times the size of the greatest bandwidth in use. For example, if a leased line is rated at 9.6 Kbps, the test file will be 10 times that amount, 96 Kb.

- Cables, connectors, and wiring to connect all the above—except the stopwatch—to any point on the network.

- Locally produced forms to record test data as required.

Nice to Have

- A time domain reflectometer to check backbone cables: both copper and fiber.

- Under certain conditions, a portable oscilloscope with a bandwidth equal to the backbone bandwidth.

- A protocol analyzer similar to Network General's Sniffer or Hewlett-Packard's (HP's) model 4972A.

- If possible, an implementation of a major network management package running SNMP or SNMPv2 such as HP's *OpenView* or Cabletron's *Spectrum*.

In the testing procedures detailed below, certain additional test equipment is mentioned. This equipment is not mandatory and may be replaced with the equipment listed above.

Testing at the physical layer is the most time-consuming and labor-intensive. In counterpoint, it is straightforward in that errors are easy to find, often visible to the naked eye. We will examine testing on the wire and cable necessary to connect LAN and WAN devices. In the following discussion, please refer to the testing diagram shown in Figure 3-7.

Physical Layer Testing

Testing LANs A and C. Several tests are made on the coaxial cable linking the PCs noted at both locations shown in Figure 3-7. These two networks are 10Base2, thin coaxial cable. They are connected to the back of each computer or terminal server by a T connector. Each end of the coaxial cable has a terminating resistor providing a 50-ohm resistance termination to the cable. Each section of the cable should be tested with an ohmmeter before being connected. Measure from the center conductor to the outer shield. The test should read "open" or "infinity" depending upon the instrument. Now measure from one end of the cable to the other end first on center conductor then shield. Both should show a very low number, 1 to 2 ohms, or "short" again, depending upon the instrument.

Repeat these same tests after the connectors have been fastened to

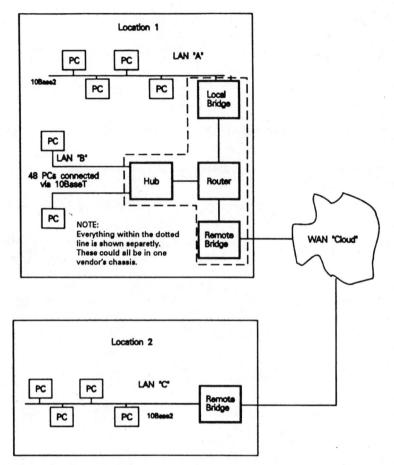

Figure 3-7. Newly installed network.

the coaxial cable. The readings should be the same. If not, the connector was installed improperly. Then measure the resistance of the terminating resistors attached to the coaxial cable. A reading of zero or infinity indicates a bad terminator and it should be replaced. The reading should be about 50 ohms, depending upon the vendor.

Unshielded twisted pair that links each PC to the hub should be tested with something equivalent to Mod-Tap's SLT3 Tester multipair testing unit. This device tests all connectors and cables for opens, conductor-to-conductor shorts, and pair transpositions. This can be done with an ohmmeter, but it will be very time-consuming.

Network Layer Testing

Depending upon the type of wiring used between the hubs, bridges, and routers, many of the same type of tests must be performed. If either coaxial cable or twisted-pair wiring is used, test procedures are the same. If fiber-optic cable is used to link these devices, then the cable should be tested with an optical time domain reflectometer or power meter to ensure the signal loss is within design parameters. Normally this is about 3.5 dB/km plus 0.6 dB/connector pair. DO NOT look into the test instrument or cable when it is being tested as some test instruments use lasers as a light source, which can cause severe retinal damage. Also, test the wiring leading to the demarcation point where the WAN circuit will terminate from the commercial carrier. Again, check for opens, shorts, and pair transpositions.

Depending upon the telecommunications software in use, the tests at the network layer may be very limited. But, in general, there are two tests: a test for presence and a file transfer test. For the purpose of this book we will assume TCP/IP is the protocol suite in use. Other protocols have similar command sets. These tests must be applied to all three networks before the bridges and routers are attached.

The ping test sends out an ICMP packet with a specific purpose, to query a device on the network with a specific address. Connect the laptop to the LAN under test at any connector. The syntax is <PING ADDRESS> where address is the unique address of the target device. Assume one of the PCs in a LAN has an address of 152.033.121.002. The syntax is then <PING 152.033.121.002>. Depending upon the application in use, the target machine will return some type of response to the laptop.

As noted above, there should be a file created for the specific purpose of testing transfer times across the LAN. Assuming a 2 Mbps throughput, we must have enough data in the file to transmit for about ten seconds. This means the file must be about 20 Mbits, not Mbytes, in size. Open a connection to the target machine. The syntax for file transfer should be something like <PUT FILENAME.XXX>. When the "enter" key is pressed, trip the starting button on the stopwatch. When the prompt returns to the screen, trip it again. The value noted on the stopwatch will be 10 times the network throughput in bps. That is, if 20 Mbits were transmitted through the network in 10 seconds, the throughput is 2 Mbps. Some versions of TCP/IP will actually give the throughput on such file transfers. Do not forget to compute the packet overhead for such transmission when figuring the throughput.

Transport Layer Testing

Certain preparatory steps must be taken before transport layer testing can begin. The first of these is hardware preparation. Many hubs, bridges, and routers can be configured while they are attached to the network from somewhere within the network. This is particularly true if a major network management software/hardware application is present. For this testing procedure we do not recommend the in-network set-up and testing process; use a laptop PC and terminal emulation software to set up these devices prior to connection. Another note: Many of these inter-networking pieces of hardware have built-in loop-back capabilities where they transmit information to themselves for a self-test. If so, make use of it to test for functionality. When testing and configuration are complete, connect to the LANs in use.

The hardware necessary to test WAN circuits is complex to operate and quite expensive. We do not recommend purchasing such hardware unless multiple WAN circuits are in use with a history of questionable quality. Once the carrier has installed the circuit and tested it, it is normally a safe assumption that the circuit is good.

As of this point in the testing, we know that all devices on each of the three LANs are functional. Ping and file transfer tests have shown the backbone and PCs are performing as they should. Now we will test for LAN-to-LAN connectivity the same way. Transmit a ping test from some station on LAN A to a station on LAN B. If it is successful, repeat the test from LAN A to LAN C. This is only one station to one station. It is not necessary to perform this same test from all stations to all stations on each LAN. When complete, the test will have completed a ping between two stations, from LAN A to LANs B and C, then from LAN B to LANs A and C, and finally from LAN C to LANs A and B. File transfer tests can be completed following the same sequence. Do not forget that the test file size should be changed to reflect the change in the throughput of the WAN circuit.

Testing at the application layer is quite difficult to quantify without having detailed knowledge of the software in use. A test may be something like the following. (Please review Figure 3-7 at this point.)

Application Layer Testing

A user at LAN A is preparing a document in a word processing package. The user requires database information in a file on LAN C. The user opens a connection across the network, queries the database, then copies the resulting reply into the document which is open on the machine they are using. Another test is to have this same user finish the document and then send it to a printer which is on LAN B.

Testing Considerations

There are several other considerations to be kept in mind while performing these tests. If the vendor (installer) tests the new hardware, software, and circuits, just how honest is the test? This vendor test can be performed when objective values have been met in terms of throughput, response time, or other quantifiable data. Someone other than the vendor should observe the tests and note the findings.

If the user tests the new hardware, software, and circuits, just how objective is the tester? Many vendors are rightfully concerned about the ability of users to evaluate new hardware and software objectively. What may appear to the user to be a defective item may, in fact, be user ignorance.

During the testing procedure certain data were gathered concerning throughput and functionality. This should be retained as a baseline to evaluate future performance as more and more subscribers attach to the networks.

This testing procedure is applicable to most new or expanded networks. It is also a high-level overview of testing and the procedures used in determining the quality of the new, or newly installed network.

Summary

Now, a review of this chapter is appropriate as quite a bit of material has been covered. The purpose of the design is the creation of the RFQ. Primary design criteria include throughput in bits per second. Other design criteria are listed in the outline below:

1. Physical distances involved both within a given segment, within a LAN and from LAN to LAN through the WAN.
2. The choice of transmission media includes:

   a. Copper wire in either twisted pair or coaxial cable.
   b. Fiber-optic cable in two general descriptions, multi and single mode.
   c. Radio in both directional and broadcast mode.
   d. Free space light.

3. After choosing the transmission media, the Network Manager must select the topology of the network. The more common choices include:

   a. Proprietary versus nonproprietary topologies.
   b. Open system topologies.
   c. IEEE-based offerings.
   d. Network operating systems have some impact on this decision.

4. Network extension hardware consists of four separate types of devices, including:

   a. Repeaters.
   b. Bridges, both local and remote.
   c. Routers.
   d. Gateways.

5. Finally, there must be some type of test plan in place for the finished product. Before the testing can be completed, the Network Manager must procure and install the hardware, software, and firmware. This material is described in the RFQ itself, which is the subject of the next chapter.

4
Request for Quotation

An elephant: A mouse built to government specifications
ROBERT A. HEINLEIN
The Notebooks of Lazarus Long (1978)

As we have mentioned previously, LAN procurement is completed in four steps. At this point the design and specification are complete. The Network Manager knows what is needed for hardware and software and must translate this knowledge into a request for quotation (RFQ). The RFQ should be understandable to all involved in the project and should be legally precise and technically unambiguous. The RFQ should also be structured so that responses to it are easily quantified.

Where Does the RFQ Come From?

The RFQ is the result of the specification and design process. Here the Network Manager puts on paper the results of the sweat, effort, and negotiations which have gone before. We will begin our analysis of the RFQ by discussing who will need to access the facts noted in the RFQ. We will then discuss the contractual elements of the RFQ. The need for objectivity and the ways in which an RFQ provides guidance. We will then evaluate the contents of an RFQ, doing a step-by-step analysis of the sample RFQ contained in Appendix B. Chapter 4 ends with a discussion of the procedures to be followed when changes to the RFQ are required.

The RFQ Is a Legal Document

The very first thing the Network Manager must realize is that an RFQ is a legal document. As we are not authorized to practice law in any jurisdiction, we must stress that the sample RFQ provided in Appendix B may not be legally sufficient in the Network Manager's jurisdiction. The Network Manager may wish to obtain legal approval either through internal legal assets or through independent practitioners. The local Bar Association can provide referral services in most states. Much the same can be said concerning the sample contract, which may or may not run in parallel with the RFQ. See Appendix C, with the same caveats concerning securing legal advice.

Legal Practitioners

Normally the RFQ and the contract are interrelated inasmuch as they refer to each other. The Network Manager must work with legal practitioners in the wording and interrelationship of these two documents.

An important criterion for the RFQ is that it must be objective in nature. Terms such as "industry standards of workmanship" or "common business practices" must be avoided. Whenever possible the RFQ must point to standards that have been set by unbiased third parties. A few examples in the following sections may be sufficient to point out some of the problems which can arise.

Specifications Must Be Quantifiable. One of the many aspects of design is the type of cable used to carry signals. Let us assume part of the network will use thin coaxial cable in the 10Base2 topology. The "quick and dirty" approach would be something like

> Coaxial cable will be functionally equivalent to 10Base2 cable in common usage.

A partially acceptable shortcut to this may be something like

> Coaxial cable for 10Base2 installations will be the equivalent of

Table 4-1. Sample Coaxial Cable Specifications

Coaxial cable to be used will meet the following specifications:

Item	Specification
Dielectric material	Solid FEP
Outer jacket	PVC or plenum-rated depending on run location
Inner conductor	Solid copper
Outer conductor	Tinned, stranded copper
Cable type	RG58U acceptable, RG58A/U preferred
Nominal OD	0.178 inches
Impedance (ohms)	50 ± 0.01
Capacitance (pF/Ft)	29.0 ± 0.05

ABC's part number XX-123 cable in both plenum and non-plenum varieties.

The primary drawback to this approach is that if this specification is ever used in a court of law, both parties may have problems defining exactly what ABC's part number XX-123 is. The preferred approach is provided in Table 4-1.

Specifications Must Point to Standards. Another example of the need for objective language may come from the building codes portion of the RFQ. The less favored approach would be

All installations will meet appropriate fire and safety codes.

Oh? Whose codes are being referenced here? Regard the following sample, then compare.

All installations will meet city, county, state, and federal standards for installations of telecommunications wiring. Where conflict between different codes arise, the installer will obtain guidance from local code enforcement officers before beginning the installation. If the installation is not treated in any codes, the installer will look to industry standards, such as EIA/TIA 568 or EIA/TIA 569, for general guidance.

The RFQ Provides Guidance

Beside being general in nature, the RFQ must provide guidance in several areas. What follows is a high-level view of five of these areas in order to set the stage for discussing the contents of the RFQ.

What Is to Be Delivered

What is to be delivered may seem to be straightforward—a LAN. But what makes up the LAN? By now the Network Manager has a complete list of cable, bus interface units, wiring hubs, patch panels, and punch-down blocks. Bridges, routers, and gateways have been specified. Does the Network Manager expect the same organization that strings the cable to also install bus interface units into the computers? Does the same person who installs the cable test it? (This last question is one which can cause serious problems and will be addressed further in Chapter 8.) Can the same firm which runs copper cable run fiber-optic cable?

Another part of the process to consider: Is it cheaper for the Network Manager to buy cable from vendor A, racks from vendor B, bus interface units from vendor C, and punch-down blocks from vendor D; or is it cheaper for the respondent to make all of the purchases? Before answering the question, consider these two scenarios.

Scenario 1. The Network Manager has bought, through the lowest bid, all of the new LAN materials from separate suppliers. When the final test procedures are applied, strange, inconclusive results follow. The installer says, "Nothing wrong with my workmanship. Must be the cable itself. You bought it!" The cable vendor says, "Nothing wrong with my cable, it's just what you specified. Sloppy workmanship in the installation."

Scenario 2. The Network Manager has one person provide both the labor and material. When the final test procedures are applied, strange, inconclusive results follow. The installer says "Well, looks like I've got a problem here. I'll let you know when it's fixed."

Which scenario do you think the Network Manager would prefer?

This does not mean, however, that the Network Manager must put all the eggs in one basket. (Chapter 7 goes into these items in much greater detail.) Still Network Managers must do their homework to ensure the situation in Scenario One never plays itself out. Consider the following alternative to Scenario One. The installer says, "Nothing wrong with my workmanship. Must be the cable itself. You bought it!" The Network Manager then shows the installer a certificate saying the delivered cable fully meets a certain set of objective specifications as of the date of delivery. The installer is charged with care of the material and so must assume the responsibility, rather than the Network Manager. Or consider, the Network Manager may want the installer to test each reel of cable before it is installed. This prevents the installer from blaming anyone else.

When Is It to Be Delivered

Once we have identified what is to be delivered, we must determine when it is to be delivered. This is not quite as simple as it sounds. In even an average LAN installation there is a considerable amount of hardware, supplies, raw materials, and tools used in the process. Where are these goods to be stored when not in use? The RFQ must include statements concerning the treatment of these items.

1. All material, including the LAN itself belongs to the respondent until accepted by the buyer.

2. The respondent will include sufficient insurance on materials and supplies to cover losses occurring before acceptance by the buyer.

3. The respondent will ensure only a minimum of materials are staged at the workplace. If fiber-optic cable is to be run inside a corrugated plastic sheath, this sheath comes on wooden reels five feet high and two plus feet wide. How many reels can be stored in an office environment waiting to be used?

4. The respondent will be responsible for removing all packing materials at no cost to the buyer.

5. Tools, test equipment, and other pilferable items are the responsibility of the contractor, and they must secure these items to prevent loss through negligence or theft. Some tools (cartridge-powered ramsets, for instance) can be dangerous to curious staffers. These and the cartridges used in them must be kept out of reach of the idly curious. Extension cords must be in good condition, with no exposed conductors.

Quality of Materials to Be Used

Thirdly, the Network Manager must be concerned with the quality of the materials used. Referring to the specification of the cable mentioned above, note that the outer jacket is PVC. How does the Network Manager know whether this is good quality PVC? It is difficult to do this without specifying a particular vendor's product line. By knowing the quality of the product, the Network Manager is restricting the chances of obtaining the best pricing. ABC cable company will charge pretty much the same price to all its wholesalers, who will mark up the price they pass on to retailers. The only difference (savings) the Network Manager can anticipate is the profit margin of the wholesalers and retailers, which is probably very little across the board.

All is not lost however. Some cable, like other electronic components, comes in three grades: residential, industrial, and military (or MILSPEC). Residential is intended for light usage only—just what the name implies—within the home or office. Industrial grade is the preferred grade for hazardous installations. This includes the factory floor, warehousing, airborne runs, or underground burial. MILSPEC cable has very little usage in most organizations. The cost differential between MILSPEC and industrial grade is quite large, and most organizations cannot justify the need for MILSPEC cable in routine installations. If the installation is shipboard or is exposed to very harsh environments, MILSPEC might be necessary. Residential grade cable can be used in some business locations, but all in all, we prefer at least industrial grades of cable for most installations.

Quantity of Material to Be Used

The next point to consider is that the quantity of material will always be more than is estimated. Let us assume that the Network Manager has an accurate set of blueprints (not usually), an accurate measuring instrument (probably), and that the cable will run where the pencil draws the line (ha!). From this the Network Manager figures 18,761.76 feet of cable will be needed. Assuming the nominal reel of cable is 1000 feet, the Network Manager will need 18.76176 reels of cable. While rounding this up to 19 reels of cable would seem to be common sense, we would recommend 20, and perhaps even 22 would not be too extreme.

Cable gets chafed, broken, shorted, or otherwise damaged during installation. The run that should consume 187.91 feet may wind up consuming three times this amount because the installers may destroy the cable in the first two attempts. The same will happen with connectors. This is particularly true with the RJ series of unshielded twisted-pair connectors and most fiber-optic connectors. Once installed, these cannot be disconnected and reconnected if done wrong the first time. Finally the Network Manager must realize that some of the material will be damaged in shipment or in storage awaiting installation. Shrinkage occurs when a worker realizes the four pair cable, connectors, and receptacles would make a good stereo speaker installation for a new apartment.

Keeping excess material after the installation is not a bad idea either. Often vendors will not sell just eight connectors, or 75 feet of cable. Once the installation is complete, there will always be a need for "just one more outlet here" or "just a pair of fiber-optic strands" there from those who did not fully consider their original requirements.

Unit Pricing

Finally the Network Manager must be concerned with the unit pricing of items within the RFQ. If, for instance, the design changes, the Network Manager must expect the price to decrease if the size of the design is made smaller, or to increase if the design is made larger. Unit pricing should consist of, at a minimum, three costs.

1. The cost of the thing itself, e.g., one RJ-45 connector costs $0.23
2. The cost of installing the thing itself
3. The cost of testing the thing

Obviously there are some things that are not tested. What is the test of a properly installed equipment rack? It does not fall over when equipment is installed. Still, it may take 1.35 man-hours at $17.85 per hour of labor to install it. These costs must be required in the RFQ and compared between respondents.

Contents of the RFQ

Now that we have a basic understanding of the items the RFQ addresses, we will discuss the more important content of the RFQ itself. The remainder of this chapter and Chapter 7, plus the RFQ in Appendix B, are closely interrelated. The reader should be aware of these interrelationships.

Administrative Information

The first part of the RFQ is the cover page(s) and the definition of terms. The definition of terms is the trickiest part of the specification as many respondents may have their definition of terms, a definition which does not agree with the usage in the RFQ.

The "Bid Instructions" portion of the RFQ is the administrative trivia necessary to guide the respondents in creating a correct and uniform response. The Bid Instructions should contain the following:

1. General, very high-level statements concerning the reason and content of the RFQ. These will be used to separate this RFQ from many others the respondents will be working on simultaneously. Scheduling and inquiries also should be part of this section.

2. A separate section will address the disposition of the responses by the author of the RFQ. Although it is considered the civil thing to do,

there is no requirement for the Network Manager to acknowledge receipt. The respondent would be wise to either deliver the response in person or through a delivery system that provides a receipt to prove delivery.

The "Basis for Award" portion of the RFQ provides the scoring system to be used by the Network Manager. Some schools of thought indicate if respondents know what they are being scored on, they will emphasize those points to the detriment of others. For example, if a respondent knows a response will be weighted because of low cost, he may ignore or overlook safety concerns in order to provide the desired response. In counterpoint, if the respondent expects each facet of a response to be treated equally, they may overlook, or at least downplay, significant items. Essentially there is no right answer to the question Should I tell the respondents what I will score them on?

Informal notification of award is the highlight of the process. To the detriment of the Network Manager, there is normally only one winner, and the Network Manager must notify the losers of their status as well. At this point, the Network Manager may discover aspects of his or her character which were hidden until now.

There are other miscellaneous areas within Section I (see Appendix B) of the RFQ which should be discussed.

1. *Duration of Quotation.* This must be some reasonable period of time—the longer the better for the Network Manager, the shorter the better for the respondent.

2. *Slippage of Schedule.* Since occasionally schedules slip, a smart respondent will build in slippage; a smart Network Manager will build in some as well. The key point is that the respondent must not hide slippage but report it when it becomes obvious.

3. *Laws, Ordinances and Codes.* List all which apply to the entire RFQ and how they will be interpreted.

4. *Technical Elegance.* An RFQ is the basis for a technical project. Theoretically, the better the technology used, the better the final product. The Network Manager should keep this in mind.

5. *Other Terms and Conditions.* This is a catch-all section for anything that does not fit anywhere else.

6. *Reliance.* This is a term of legal art, which means what lawyers say it means. By including it in the RFQ, the Network Manager provides lawyers a sense of how the Network Manager views the respondents.

Broad Description

Section II of the RFQ is free form. This allows the Network Manager to wax poetic in the description of the final product. The only thing to be concerned with here is that the poetry does not conflict with the prose in other sections. In all seriousness, this section provides a good point for a general discussion of what the final product will be able to do without getting into technical detail. That follows in Section III. One concept should be mentioned here. When and if differences arise between the respondent and BMC, many attorneys look for some evidence of the intent of the parties. *Intent* is a legal term of art under these conditions. The Network Manager should cast some part of Section II to show BMC's intent in this matter.

Hardware and Software Specifications

Section III of the RFQ lists specific details about the hardware, firmware, and software, which will be used to make up the new network. First we will discuss several items found in many modern networks. Besides the devices noted here, there should be specifications for file servers, workstations, patch panels, equipment racks, network operating systems, utility software, backup media, and a plethora of other devices that are not shown. After discussing fiber-optic cable for harsh conditions, remote bridges, and uninterruptible power supplies we will go into some of the minor details finishing up Section III.

Section III, paragraph D addresses the installation of fiber-optic cable but does not specify the cable itself. That is not an oversight, such specifications will vary from time to time and from place to place. Table 4-2 provides a sample specification of six-strand fiber-optic cable, which can be installed in wet areas suffering a high range of temperature variation.

In counterpoint, note that Section III, paragraph C does not specify any technical characteristics at all. Again, this is not an oversight. That paragraph points to a specification that is available to all persons. There is no reason to duplicate the criteria in a previously published document.

Also within Section III should be the specifications for other types of equipment. Table 4-3 shows the specification for a remote bridge used to tie together the two locations BMC occupies, 10 and 14 Industrial Park Way.

Bridges which are used to link networks together require constant power from standby power supplies of some sort. Table 4-4 provides a specification for an uninterruptible power supply useful for the

Table 4-2. Specification for Fiber-Optic
Cable under Harsh Conditions

Category	Specification
Strand count	6
Mode	Multimode
Diameter (inner/outer) MM	62.5/125
Bandwidth	400 MHz
Filling	Gel filled, loose tube
Strength member	1+
Inner shield	Kevlar
Outer shield	Polyethylene
Temperature range	−40 to +60°C
Humidity	100%
Weight (kg/km)	36

Table 4-3. Specification for Remote Bridge

Category	Specification
Filtering in packets per second	No less than 14,400 64-byte packets
Forwarding in packets per second	No less than 10,750 64-byte packets
Dimensions (for standard 19" rack)	19"W×4.5"H×8"D
Power requirements	110 VAC ± 5%, 3 VA, 50–60Hz
Protocols supported	IEEE 802.3 (DIX version), TCP/IP
Network manager compliance	SNMPv2
Number of ports	96
Expandable	No
Hot-swappable PC boards	Yes
Power supplies	Dual, shared load
Administration	Remote (in band) local (out of band)
WAN output	DS1 or subrate
CSU/DSU	Yes (internal preferred)
Connectors	96 RJ-45 female, 1 RS-232 (for local administration), 1 V.35
Quantity required each location	2 (total of 4)

bridges shown in Table 4-3. Note also that, if necessary, this power supply could be used as an uninterruptible power supply for user workstations. If it is to be used to power the file servers, a much higher rating is needed in the KVA and operating period categories.

Section III, paragraph F is vitally important for several reasons. Let us examine each of these.

During the installation of fiber-optic cable, the strands will be prepared for use. It may make more financial sense to only prepare those

Table 4-4. Specification for Uninterruptible Power Supply

Category	Specification
Power rating in KVA	3
Input voltage (AC)	115 or 230
Input frequency (Hz)	50 to 60
Harmonic distortion	Less than 3% THD
Power factor (lead/lag)	0.67
Internal battery	Lead/acid, sealed, maintenance free
Operating period full load	16 min
Dimensions (approx)	38"W×20"H×28"D
Weight in pounds (approx)	60
Operating temperature range (C)	0 to 55
Humidity range noncondensing	95%
Heat dissipation (BTU) (approx)	1100
Quantity	1 per workstation, 1 per 2 bridges

that are to be used during the initial phases, but this is not so. Consider that to bring the installers back will cost some measurable amount of money. Then, the second or subsequent cable preparation must be tested, which again costs some amount of money. In almost all cases it is preferable to have all the fiber-optic cable prepared, connectorized, and tested at the same time. As an aside, using this method makes connecting new equipment much faster and cheaper.

In Section II, paragraph F the sub-paragraph stating, "Respondent will base their response on the following factors in their given order of precedence. Human safety, reliability, expansibility, flexibility, telecommunications security, space utilization, and human factors" is key when the respondent will be the one designing the wiring closets or rooms. We have seen the time when technicians were forced to climb cable trays—a practice not recommended—to add or remove a cable. This activity is not only dangerous; it is stupid. Under the concept of reliability, the Network Manager should be concerned with robustness. Will the cable tray fall from the ceiling? Are the equipment racks securely bolted to the floor, and where possible to the walls? Expansibility includes things such as sufficient power outlets for rack-mounted equipment, or floor space for additional racks. Cable trays must be large enough to accept at least a 100 percent increase in loading before replacement. The concept of flexibility is applicable to changes. Is there space to move equipment racks and cable trays if necessary? Will the floor loading accept new configurations? Part of telecommunications security is physical security. Does the design allow for unauthorized access to the wiring closet? Are there several doors or windows where people may enter without being noticed? Finally the concept of human factors must be considered. Is there room enough behind the equipment racks for technicians to work with some degree of comfort and convenience? Is there enough light and ventilation?

In keeping with the points above, the Network Manager cannot merely pass the responsibility of the room design to the respondent. The Network Manager must evaluate responses and select the best one. This is the reason for sub-paragraph 3, paragraph F stating: "Layout, including rack locations, will be submitted for approval prior to the start of installation."

No LAN can be installed without some type of numbering system. This system assigns a unique number to every run of fiber-optic or copper cable used in the LAN. Numbering systems are one of those things which at first may appear simple and clear, yet in the end can cause more problems than a much more complex process. Although there are many ways in which to number cables there are only two worth using. We have, in the RFQ, demanded the derivative approach (see Figure 4-1).

Assume we identify a hub as OPS_001. That could stand for the first hub to be installed in the operations division of BMC. The cable which connects from a supervisor's computer to the hub would have the number OPS_001/A. The next cable connecting to the hub, say from Mary Woods's computer, would be OPS_001/B; and so forth. Now, if there are multiple hubs which, in turn, are connected to a local router, then the same principle applies. Assume a router has been given an identity of ROUT_001. The cable connecting the router to the third hub would be ROUT_001–OPS_003/A. Let us assume there is one high-level router at 14 Industrial Park Way and another at 10 Industrial

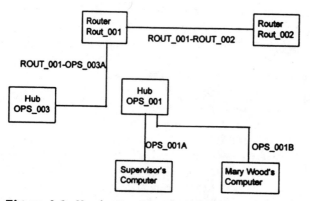

Figure 4-1. Numbering scheme using derivative numbering.

Park Way, and they are connected by a high bandwidth circuit. The number connecting a high-level router to one of the lower level routers would be numbered ROUT_001_ROUT_002. It is obvious through inspection that such a numbering scheme is extensible and easy to read and to understand. It fits well into most commercial databases or spreadsheets for manipulation.

The other approach to numbering is that of geographic coordinates. Note in Figure 4-2 we are dealing with a rectangular building.

Lay a grid across the building with letters on one axis, numbers on the other. These letter–number combinations mark off one square. Within that square are numbered drops or connections. Square C2 has nine drops within its borders. Each cable is then given a number somewhere between C2-1 and C2-9. Add another cable run? Simple, just add another number. If the building has multiple floors, then add another digit to the grid coordinates. The first floor would be 1C2, the second would be 2C2 and so on. If the building is not totally square or rectangular, then let the grid be laid on areas which will not be used as well. See Figure 4-3 as an example.

This approach has several drawbacks. First is the ability to judge the size of the grid. If the grid is too small, only one or two outlets will be used. This is getting too close to one unique number per grid, a situa-

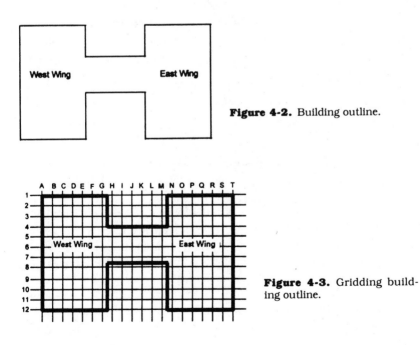

Figure 4-2. Building outline.

Figure 4-3. Gridding building outline.

tion to be avoided at all costs. Too large a grid and there are too many outlets, again a situation to be avoided. As a rule of thumb, there should be between five to nine outlets per grid square. If office sizes are permanently fixed, then the grid should be disregarded and office numbers used. Still, this approach does not fit the majority of modern offices or manufacturing plants. There is one other drawback when using the geographic approach. When using derivative numbering systems the technician can, after a while, know which person is connected to which device. But with a geographic approach, a drawing of the building is required to know where cable 2C2-8 is terminating.

Here we see that Section III is the key part of the RFQ, addressing actual hardware, firmware, and software the Network Manager has selected. These descriptions and specifications are the result of long hours of study and effort by the Network Manager, the staff, and perhaps one or more consultants. This is not the end however. Now the Network Manager must worry about the actual installation of the selected components.

Installation

Section IV covers the installation and the relationships between the installers and BMC in the person of the Network Manager. BMC has certain responsibilities which must be met to help the respondent accomplish the installation. For instance, if the contractor uses power tools to drill holes or cut conduit, he must know whether commercial power is available or whether to bring a power generator. We will review some of the core requirements of BMC under the terms of this RFQ. In doing so we will not reiterate the information contained in Appendix B, Section IV, but consider the following:

"BMC will provide a safe, clean workspace." If a major building renovation is going on during the installation or the installation parallels the construction of a new building, there are always hazards. BMC must ensure the respondent's employees are not unnecessarily exposed to danger. At a minimum, BMC must make the respondent aware of potential hazards during the bidding process.

To successfully prepare an honest and reasonably accurate bid, BMC must provide the respondent with scaled, accurate floor plans of the building(s) receiving the new hardware. The respondent must know how many feet of wire it will take to connect an outlet to a patch panel. The respondent must know where structural walls are versus those walls which merely divide one room from another. The respondent must know where cables enter the building and where there are raised floors or dropped ceilings. This requirement for floor plans is not all one way, however. The competent, honest respondent will provide BMC with the same set of drawings, or another just like them, with the routing for the new cables and the location of equipment rooms/racks shown. This is, in the trade, known as "redlines."

The third important part of BMC's responsibility is to promptly test the equipment as it is installed. Often parts of the installation are finished ahead of others and work must halt until these completed parts are accepted. BMC owes the respondent the courtesy of testing and accepting or rejecting the completed parts when the respondent has completed a phase of the effort. Such a requirement should be noted in the project management schedule pertinent to the installation.

Besides the obligations from BMC to the respondent, the respondent has additional obligations to BMC. The respondent has access to many parts of BMC's property and must take exceptional care of the property. Again, it is not the intent to fully duplicate the material in Appendix B, but just to highlight some of the key points.

Major installations require large amounts of hardware. Where is this material to be stored during the installation? The respondent owns it until BMC provides final acceptance. It will be to the respondent's advantage to store the material on BMC's property. But this raises certain legal questions BMC should pursue with a practicing attorney. As a general rule, we recommend that material not be stored on BMC property at all, under most conditions. If such storage is required, then BMC must get a waiver of responsibility from the respondent.

The respondent and BMC both must realize that during construction, accidents can happen which may damage BMC's property. While the respondent has insurance to cover these accidents, the respondent also has a responsibility to ensure the workers are well trained in accident prevention and know the proper procedures to follow in case of accident and/or personal injury. The respondent is also responsible for ensuring that the workers under his control have proper safety gear including hard hats, safety glasses, breathing filters, and safety belts if required.

A key point the respondent must keep in mind is that the final installation must pass inspection by codes enforcement officials. As this varies from town to town, even within one state, the responsibility rests on the shoulders of the respondent. BMC has very little control over this matter so should be out of the way. Of course, the language of the contract indicates the respondent will not be paid until the installation meets or exceeds the appropriate codes and ordinances.

Besides insurance covering the employees, the respondent must maintain insurance on the material being installed. If, during the installation process, a fire destroys expensive cable and electronics, the respondent "owns" it. This insurance will reimburse the respondent and the work will go forward immediately. BMC may, at its discretion, ask to be provided with copies of the policy covering such accidents. BMC should not allow the respondent to be self-insured, which is a means of working around the insurance requirements. If the respondent refuses to provide such insurance, BMC is recommended to purchase it with the understanding that the respondent will have the cost of the insurance deducted from the total payment.

In large installations, many respondents contract with other organizations to perform certain services which the respondent may not feel competent to provide. These subcontractors must be bound by the same rules and requirements which bind the respondent, and BMC may or may not require the respondent to provide documentary evidence attesting to this fact.

Test and Acceptance

Section IV, paragraph 14, parts C and D discuss the testing procedures to be used for the final product, the LAN. Before getting into the technical aspects of the tests, the respondent must acknowledge, in writing, that they will comply with the testing provisions of the RFQ. Failure to comply with this provision will be grounds for summary dismissal from consideration.

In general, tests are conducted to measure compliance with objective, agreed-upon specifications. In other words, the bridge will forward X many packets per second, assuming a packet size of Z bytes. The tests to be used are not graded on a percentile, rather they are on a go–no go basis. The Network Manager has the responsibility to design the test, observe it, and record the pass or fail status. Here the Network Manager may wish to seriously consider hiring a consultant to create and observe the tests. A consultant is an unbiased third party, one whose specialty complements the general knowledge of the Network Manager. A more accurate, and correspondingly more expensive approach is to hire engineers from the manufacturer of the product being installed to design and observe the test. However, we cannot think of test conditions outside those used for life support or for Department of Defense applications which require this degree of accuracy.

How many of the products to be tested must actually be tested? Assume a LAN has 30 cable runs: Shall all 30 be tested? The obvious answer is yes. What if the number is 300? Can we define a statistically significant sample for testing, or do all 300 runs become subject for testing? What if the number of runs is 3000? Testing 30 runs takes maybe 90 min with a sharp crew and minimum repairs. Testing 300 runs will take significantly longer. In essence there is no textbook solution to testing a statistically significant sample. We do not recommend a sample-based approach, but the final decision rests on the shoulders of the Network Manager.

Like anything else, the work is not complete until the paperwork is done. The paperwork concerning testing is vitally important if there is any argument over whether the respondent completed the job on time, as specified. Test forms must be agreed upon by the respondent. The results recorded on those forms must be dated. The people conducting and observing the tests must be identified by name. Requiring signatures of technicians and engineers is not unheard of. These forms are the only objective proof the Network Manager has that the LAN operates as specified. Although respondents may argue the fact, they should realize such tests work in their favor as well.

This discussion completes most of the technical aspects of the RFQ, although there are three other areas that must be touched upon before ending this particular section.

Qualifications of Bidder's Staff

As was asked of us one day, "How high is the high school?" The quality of the respondent's effort is controlled largely by the quality of the people who perform the work. This, in turn, is controlled by their training and education. Respondents, if they are proud of their work force, will have no objection to providing documentation of these people's training and education. The Network Manager must make the effort to determine if the documents are accurate and sufficient for the job under consideration.

Financial Qualification of Bidders

The respondent also must provide evidence of financial stability. An annual report to the stockholders is not sufficient since such reports are as flexible as a rubber ruler and prove nothing in some cases. Instead an independent financial evaluation of the respondent is recommended. These evaluations are all part of the cost of doing business, and the respondent who refuses to provide copies of such evaluations should have many other positive things going for him or her before the Network Manager accepts the firm as a finalist in the decision-making process. A firm teetering on the brink of bankruptcy is not a prime candidate for the installation of a new LAN since it may not be in business long enough to finish the work!

Acceptance Form

The acceptance form is the only document which the respondent is asked to sign that the Network Manager can control. It is the most vital element of the response to the RFQ. This document binds the respondent to the terms and conditions of the response and must be dated and signed by a person empowered to bind the respondent to these terms and conditions. Whereas there is no legal requirement for it, the Network Manager may require the signer to have the signature(s) notarized. Again, the Network Manager should consult with a legal practitioner in this area.

Appendixes and Annexes

The RFQ may require appendixes covering anything from blueprints to manufacturer's specification sheets. These should be included as separately lettered appendixes to the RFQ and should be listed in the table of contents for respondents' use. We have found that respondents also add appendixes to their responses and to eliminate confusion between the Network Manager's and the respondent's appendixes, we recommend that the respondent number rather than letter their appendixes. Shown in Table 4-5 are two extracts from tables of contents for an RFQ and a response. This requirement should be spelled out in the directions in Section II of the RFQ.

Change Methodology

Despite the effort the Network Manager puts into the RFQ and how often it is reviewed by the "powers that be" or even by an outside

Table 4-5. Appendixes and Their Identification

An RFQ for a LAN for BMC	A Response to BMC's RFQ
Table of Contents	Table of Contents
Appendix A. First floor blueprints	Appendix 1. References
Appendix B. Second floor blueprints	Appendix 2. Auditor's report
etcetera	etcetera

agency, there will be questions, revisions, or changes. Where it may not be legally required to inform all respondents of these changes, the Network Manager owes them the courtesy of informing them so they can alter their responses accordingly. The Network Manager must create a mechanism for receiving questions, developing answers, and then communicating these answers to all others involved. What follows is one approach to this process.

1. Determine a date about four weeks prior to the deadline for responses. In this instance it will be 1 April.

2. Schedule a conference with all respondents sometime during that day to inform them of changes which have occurred.

3. Ensure that all questions from respondents are put in writing prior to the conference.

4. Provide the same written response to all respondents at the conference. Make every effort to mail these changes to those who are unable to attend.

Occasionally oversights occur, or business conditions change radically from the time the RFQ is released until the time of awarding of the contract. The Network Manager must prepare for these changing circumstances by having a paragraph which allows for the total withdrawal of the RFQ with no obligations on the part of BMC. The exact wording of such a paragraph should be the work of a legal practitioner.

Summary

As has been stated previously, the entire LAN procurement process is an iterative cycle. The RFQ is the third phase of the cycle and reflects the results of the specification and design stages. The RFQ codifies the specification and design through good writing. Therefore:

- The RFQ must be thought of as a contractual document.
- The RFQ must be objective in nature.
- The RFQ provides guidance to respondents and others involved in the procurement process.
- The RFQ has seven major divisions in most cases. These include

  A narrative listing the objective of the RFQ.

  The requirements of the new (or modified) system.

  Procedures used during installation, testing, and acceptance.

Some means of determining the qualification of the respondent's staff.

Guidance for the respondents in providing information concerning their financial condition.

The appropriate forms for the respondent to complete.

Appendixes to the RFQ itself.

- The RFQ must have some mechanism for communicating changes to all parties involved.

Now that the RFQ has been written and distributed, the next step is to analyze the respondents' responses.

5

Analysis of
Responses

Let blockheads read what blockheads write.
PHILIP STANHOPE, EARL OF CHESTERFIELD
Letter to His Son (1774)

This chapter addresses the simplest and yet most important aspect of the entire RFQ process: Evaluation of the responses. To provide a groundwork for this activity, we will discuss how to use objective data in the analysis, what the intermediate results will be, how to reduce the respondents to four or five semifinalists, and finally how to negotiate with these semifinalists for a final offer.

"I know what I said. Is what you heard what I meant or is what you heard what you wanted me to say?" The Network Manager must be aware that this mindset is prevalent in all parties to the RFQ process. As alluded to in Chapter 1, the Network Manager must have some way to evaluate the responses provided by the respondents and rank order them. This starts with assigning point values to the responses.

Evaluations Must Be Based on Objective Data

Perhaps the best way to describe objective data is to provide several sets of examples. Although some of these may appear obvious, the Network Manager should consider some of the subtleties involved.

Example 1

Subjective Description: The network interface card will provide throughput at 10 Mbps.

Objective Description: The network interface card will provide throughput according to the IEEE 802.3 specification for 10BaseT topology.

Discussion: We start by eliminating the obvious objections. 10BaseT does not always operate at 10 Mbps. It is specified to operate at that speed, but does it? The only way to tell is to measure each network interface card's actual operating capabilities installed in the computer. If there was ever a complaint by the respondent or the Network Manager, each card would have to be tested separately.

Now let us examine a more subtle side to this argument. Any respondent can claim a product has a throughput of 10 Mbps. Further, there may be some evidence to back this up. But will it work with other network interface cards at this speed? It is kind of hard to tell about that. It would be better if the respondent states that the product meets an identifiable, objective specification since that type of claim is much harder to fudge.

Example 2

Subjective Description: The routers used will have a throughput of 14,400 packets/sec.

Objective Description: The routers used will have a throughput of 14,400 64-byte packets per second in a 10BaseT topology.

Discussion: Here the problem is not with subtleties but with completeness. The subjective description assumes the Network Manager and the respondent are talking about the same thing, but this is not necessarily so. Usually there will be a need to add "...64-byte packets...in a 10BaseT topology" even though by the time the respondent reaches this point, it is clear that the Network Manager is talking about a LAN using a 10BaseT topology.

There are considerations when this level of specificity is required. What if someone other than the Network Manager evaluates the respondents' responses? This could happen if the Network Manager leaves or the evaluation of the responses is conducted by an outside agency or consultant.

Example 3

Subjective Description: The hardware will be tested when installed and results provided at the end of the installation period.

Objective Description: The hardware testing procedures will be specified in detail when submitting the bid.

Discussion: Here is a not-so-subtle point for consideration. It is self-evident that a device must be tested after it is installed but before the LAN is declared operational. The depth and breadth of the testing provided will point up differences in the quality of the work provided by the respondent, a key point in the Network Manager's considerations.

A more subtle point is that of the respondents' knowledge of the product line. Assume a local or remote bridge is required at some point in the LAN, perhaps one at every department. The respondent notes they will test the local bridge for its ability to filter packets. Normally such filtering is done on a forward/discard basis. That is, packets will either be forwarded through the bridge or discarded. The testing procedures will show how well the respondent understands this function.

Examples such as these are limited only by the experience of the author of the RFQ or the Network Manager. These three provide the reader with some idea of why objective data are required. Once this type of data are on hand, what does the Network Manager do? The next section shows how to use such data.

Intermediate Results

The intermediate results of the evaluation will appear in a spreadsheet. Let us take another look at Chapter 4's specification for the remote bridge, the device which is used to provide connectivity between the two buildings owned by BMC. This is shown in Table 5-1.

Table 5-2 is an extract from the spreadsheet used to evaluate the responses. Note that no respondent's name is shown in any column. This is done for several reasons. First, it makes for narrower columns, thereby making the comparison easier. Second, the use of a number rather than a name prevents the evaluator from being too biased toward or against any particular respondent. Of course, there must be a separate list that cross-references bidder numbers to names.

Next, notice there are two columns with no values noted: the item and the criteria. The column headed Item is the feature being evaluated. The Criteria are the hardware functions and information provided by the respondent. In the Value column are the points assigned by the Network Manager to the importance of each feature. Here the Network Manager finds the concept of a 110 VAC operation to be of limited value (five), whereas the number of ports is of equal value to the count of bridges being bid. Remember the cost is the cost of all bridges, not just the operational ones. The Network Manager has specified two operational bridges and two hot standby bridges.

106

Table 5-1. Specifications for Remote Bridge

Criteria	Value
Filtering in packets per second	No less than 14,400 64-byte packets
Forwarding in packets per second	No less than 10,750 64-byte packets
Dimensions (for standard 19" rack)	19"W×4.5"H×8"D
Power requirements	110 VAC ± 5%, 3 VA, 50–60Hz
Protocols supported	IEEE 802.3 (DIX version), TCP/IP
Network Manager compliance	SNMPv2
Number of ports	96
Expandable	No
Hot-swappable PC boards	Yes
Power supplies	Dual, shared load
Administration	Remote (in band) local (out of band)
WAN output	DS1 or subrate
CSU/DSU	Yes (internal preferred)
Connectors	96 RJ-45 female, 1 RS-232 (for local administration), 1 V.35
Quantity required each location	2 (total of 4)

The values assigned by the Network Manager are almost purely subjective in nature. What is important to one Network Manager is not as important to another. For instance, let us assume that the bridge is to be installed in a network that supports a mission-critical application, perhaps because production-line control information flows through it. This instance would require a bridge with a very large mean time between failure. On the other hand, a Network Manager who is more concerned with real-time network monitoring and control may put more emphasis on nonproprietary network management protocols. Here the Network Manager does not put too much emphasis on management, a value of 12, but puts a very high level of emphasis (25) on the number of connections to the bridge.

A side point may be worthy of consideration at this time. The order in which items are listed may, to some people, be important. In theory they may wish to place the most important items first, the rest in descending order of importance. We have found that this is not a particularly useful approach but does help the respondent in determining the Network Manager's priorities in a more orderly fashion.

The columns are filled out by merely going through the responses and extracting the appropriate information. Does the respondent state, unequivocally that their product will filter 14,600 64-byte packets/sec?

Table 5-2. Spreadsheet for Response Analysis

Item	Criteria	Value	Bid 1	Bid 2	Bid N
Filtering	14,400 64-byte packet	20	20	20	10
Forwarding	10,750 64-byte packet	20	8	20	10
Dimensions	19"W×4.5"H×8"D	7	7	0	7
AC power	110 VAC	5	5	5	5
Protocols	IEEE 802.3, TCP/IP	20	10	20	10
SNMP?	SNMPv2	12	12	12	0
# of ports	96	25	15	25	25
Hot swap	Yes	10	10	0	0
Pwr. supplies	Dual, shared load	8	0	8	0
Admin.	In/Out of band	10	0	10	10
WAN output	DS1 or subrate	12	12	12	12
CSU/DSU	Yes	6	0	6	6
Connectors	1 per port	20	20	20	20
Count	4	25	12	25	25
	Totals	200	131	183	140

If so, then the full 20 points is granted. What if the filtering was limited to 7300 64-byte packets/sec? Should the Network Manager grant half the possible points? In general, the answer is no. The filtering capability is a performance issue, and there is a great deal of difference, from a performance aspect, between filtering 14,600 packets and 7300 packets. We could not recommend more than three or four points for a bridge with these capabilities. The interpretation of how many points to be given for partial noncompliance with required criteria is something that is learned only through experience.

We shall not leave this area immediately. Another question is how many points are given for an item which is not quite close enough. The requirement for forwarding is 10,750 64-byte packets. This provides a respondent with a possible maximum of 20 points. If the respondent provides a device with a forwarding rate of 10,500 64-byte packets, does the respondent receive zero points? Perhaps 10 points? Would 15 be closer? Is 19 correct? This is a very subjective decision on the part of the Network Manager and tough to defend in a court of law. The only way the Network Manager can provide self-protection in this case is to have a fully documented point award structure for each category. It is a lot of work only if an award is not challenged. If a suit results from the award of a bid, the Network Manager will welcome the effort which went into such documentation.

The Network Manager may not be the only person to fill in the number of points to be awarded to the respondents. Are there others in the organization capable of interpreting the information provided? By reviewing the organization chart of BMC, we find the MIS director, a programmer, and perhaps the telecom technician also may be capable of analyzing the respondents' responses. The result will be a spreadsheet.

This process is then repeated for each item which is part of the RFQ. What is shown in Table 5-3 is a summary of several items, specifically those noted in Chapter 4, in still another spreadsheet.

Table 5-3 provides a summary of bids from all respondents for all categories. Multiple categories have been left out for brevity's sake. The value column indicates the maximum value allowed by the Network Manager for each category. It is obvious by inspection that Bid 2 is the best qualified bidder as far as the point standing is concerned. Yet there is a "rule of thumb" which states that the Network Manager should throw out the highest and lowest bidder. The highest, because the respondent may have disguised or even hidden things in their response. The lowest because the respondent is so far off the design that the firm is not worthy of consideration. This rule should not be practiced unless there are many responses.

Unit Price

We have seen how the Network Manager determines how well the respondent can meet the design requirements for the new network. This by itself is somewhat useful, but there are two other items the Network Manager must consider, the first of which is the unit price.

The unit price is the cost of an item. One RJ-45 connector costs some readily identifiable cash amount. The respondent must identify this to provide the Network Manager some idea of how much the respondent is gouging (oops), is making a profit, on the material. For the purpose of demonstration, let us assume all respondents but one charge about $0.26 per connector. One respondent charges $0.125 per connector, less than half what the others charge. Why? Are the others price gouging, or has this respondent missed the specification somehow and is using a connector which is not within specification? Or, possibly this one respondent has a large amount of connectors left over from previous jobs, has taken his profit on them, and can afford the lower cost.

One of the advantages of unit pricing is to enable the Network Manager to determine how much the cost of the new network will go

Table 5-3. Summary of Other Spreadsheets

Category	Value	Bid 1	Bid 2	Bid N
Fiber-optic cable	120	120	120	0
Remote bridge	200	131	183	140
UPS	35	35	35	35
Total	860	637	772	459

up if additional subscribers are needed. If the unit cost of a connector is $0.26, and the cost of cable is $0.83 per foot, it is easy to compute the cost of a new subscriber. Two connectors at $0.26 apiece equal $0.52, added to the cost of 176 feet of cable at $0.83 per ft ($146.08) provides a total materials cost of $146.60.

Cost of Labor

The second item is the cost of labor to install hardware. The respondent must break out these labor costs, on an hourly basis, and separate them from the total cost of the bid. Some respondents say that this is impossible to do. (Yes, well, humm, and other politically acceptable silence fillers.) From experience we know a respondent determines installation costs of cable on a per run basis. The respondent knows that X numbers of installers can run Y number of feet of cable on an hourly basis in the type of building where the new network is going. Why else will the respondent want to walk through the building if not to get a feel for the place? Paying those installers Z number of dollars per hour, the formula is quite simple. X times Y times Z equals the total cost of installing cable. Multiplying that figure by the markup, the respondent knows how much each run of cable will cost, on average, to install. The word *average* is the problem for the Network Manager. If a cable run is extra long, or requires more than "average" time to install, then the estimate for additional subscribers may be off.

Selection of Semifinalists

There is more than merely adding up a series of numbers to the process of selecting a semifinalist for the short list. The astute Network Manager will be able to evaluate many responses by the format and content of the bid. If the respondent simply fails to meet the requirements, then the bid can be discarded. If the response is padded with sales literature and very little actual information, it, too, should be a serious candidate for the rotary file.

The Network Manager should look for obvious errors such as mathematical functions. If the respondent notes 2345 RJ-45 connectors at $0.26 per unit, the total should be $609.70. A rounding up to $610.00 would be acceptable, a rounding down to $600 is very good, but a cost of half the total of $609.70 is an indicator the respondent has not done his homework. On the other hand, watch out for bids which all end in zeros. Table 5-4 provides a response to an RFQ where the following (not real) figures were noted. Which of the numbers in Table 5-4 would the reasonable Network Manager think are close to accurate? We would assume the respondent knows quite well the cost of the computers and software but is guessing the cost of labor and taxes. And we are relatively sure the respondent has not guessed in the favor of the Network Manager, and even if they did, would not accept the miscalculation. A cost overrun in the respondent's favor will appear in final accounting of this contract. The firm authoring this response did not receive the winning bid.

Other selection criteria involve the presentation of insurance or bonding instruments with the response. The RFQ and perhaps the contract itself will require the respondent to show demonstrable proof of insurance and in some instances bonds for the work to be done. Often insurance policies are only good for one year; bonds normally have an expiration date as do insurance policies but other factors can affect them. Regardless, the Network Manager must check and ensure the insurance policies will not lapse before the scheduled end of the project. We have seen several potentially winning bids thrown out because of this simple oversight. Depending upon corporate policy and the wording of the RFQ itself, the Network Manager may wish to go back to the respondent and point out this problem and allow the respondent to correct the oversight.

If the Network Manager has a respondents' conference sometime during the process of issuing the RFQ (not a bad idea in a complex project),

Table 5-4. Summary
of Cost Categories
(Example)

Labor	$23,000.00
Computers	$181,763.07
Software	$93,435.22
Taxes	$1,200.00

then the Network Manager may get a good feeling for the quality of the final product. Do the respondents' representatives ask intelligent questions or merely use this time as a showcase of their alleged capabilities? Do the representatives at least use the correct language in discussing the technology of choice? Are they professional in the treatment of their peers during this process? Anything negative between several organizations may be a trigger for the Network Manager's concern.

Project Management Plan

If the project is complex, the Network Manager must specify that the respondent provide a project management plan. There are several points to consider when evaluating this plan.

For instance, are the timelines specific and realistic? We know one very large project where the respondent's work breakdown structure indicated that it would take one day to install all equipment racks and mount the hardware in the racks. Considering there were in excess of 75 racks and the respondent allocated only three people to that task, we were somewhat confused. Contact with the respondent indicated that it should have taken 11 working days: A simple typographic error reduced the job from 11 days to one day.

Are requirements realistic? In another project, the respondent noted the requirement for a project manager's office consisting of a room 12 feet on a side, with light, heat, power, and telephone connection. This is not an unusual requirement, but considering the entire building was scheduled to be gutted before new wiring is installed, it would be extremely difficult to fulfill this requirement.

Is sufficient labor available? We remember one RFQ which would have required 35 workers on site at one time. It was known the respondent had a staff of 14. Where would the other 21 workers come from? The respondent, when questioned on this oversight, indicated that he would be providing a subcontractor for this phase. Subcontractors were not prohibited; in fact, they were specifically allowed, providing the respondent identifies them as such, an action this respondent failed to do.

One final point on the project management plan. Although it is quite difficult, the Network Manager should try to determine what assumptions the respondent is making in the plan. Are there assumptions for access to the building? Do assumptions include no delays because of weather, nondelivery of equipment, or other actions beyond the control of the respondent? We do not recommend the Network Manager probe each respondent for such assumptions, but friendly conversation can reveal much of the thinking behind the respondent's response.

The Network Manager must remember the entire effort of this process is to eliminate the "also rans," the "weak sisters," or other marginal respondents. This process is not complete without the fatal reference check. There are two questions that must be asked of any reference in this process.

> Are you satisfied with the work provided by (enter respondent's name here)?
> Would you hire them back to perform work similar to the work done before?

A resounding yes to both puts the respondent on the short list. Any other preliminary questions are superfluous and will result in useless responses.

What about answers that are not so emphatic? Many well-meaning people find it difficult to say something bad about a respondent, so they gloss over the negative aspects with statements such as "competent workmanship" or "no complaints, but we won't be installing any more in the near future so I didn't even consider it."

Bad workmanship is not necessarily the reason for a lukewarm reference. Often the original reference source has moved on, and the replacement really has no idea of the quality of work the respondent performed. Inquire if this is the case; if it is, ask to speak with someone who was there for the work. We have found the further down the company hierarchy the Network Manager asks, the more accurate—and sometimes vehement—the response will be. We have actually received responses from analysts and technicians that can be accurately paraphrased as "I wouldn't have those slobs back here on a bet!"

If the response is lukewarm, sometimes the Network Manager may wish to probe for detail. "Thank you Ms. Jones, I understand they did a workmanlike job, but what could have been done to make it better?" Or, "Yes Mr. Smith, I understand you had just been assigned the position when this was done. What changes have you had to make since the job was completed?"

Finally, check all references. If a respondent is Start-up Inc., they may not have too many, but are working hard with the few they have. Big Time Networking Company, Ltd., probably has more references than the Network Manager may be able to check. A rule of thumb may be applied here: Check until you get a bad one, or cover the last three years. The turnover in many firms may eliminate the more incompetent workers, so the references before that time are not to be trusted.

Although it is quite hard to quantify, the Network Manager may wish to consider the professionalism of the response in the paper itself.

A slick, well-bound document may hide a poor response. In counter-point, the occurrence of misspelled words may not indicate a poor response, just a hurried one. The second item to consider, which is hard to quantify, is the depth of understanding shown. Does the respondent really understand what the RFQ is supposed to do? It is easy to parrot words back to the Network Manager, but probing questions may determine the depth of the respondent's understanding. The better the respondent understands the requirement, the better the design will be.

This evaluation process can be summarized in two categories: the hard numerical databased on an evaluation of the respondents' responses and the subjective evaluations from references, discussions, and analysis of the form, not the content of the response. The result is a winnowing process, leaving the Network Manager with a few high-quality candidates. We recommend no more than five and no less than three. Experience can be the Network Manager's only guide to how many to deal with when the time comes for the respondents to provide their best and final offer.

Best and Final Offer

The best and final offer (sometimes known as BAFO) is the chance for each of the respondents to *reconsider* their mathematics, hardware, and labor calculations. Often respondents will dash off a response, not really expecting to get the bid. Low and behold they find themselves in the finalist category and now must either live with what they have written or drop out. We have seen one respondent's firm come in so low, it was on the verge of signing the contract before the firm realized what it had committed to. Needless to say, the bid was withdrawn and the purchaser had to go with the number-two choice.

Let us assume, for discussion, that the bids for BMCNet are as indicated in Table 5-5. The span between high and low is $12,695.15, or less than 2.4 percent of the highest bid. Not many deals are thrown out over that small a percentage. The question the Network Manager must ask is how do I get the best price? The answer is to dicker, bargain, and negotiate to reduce prices. The Network Manager may have a great deal of leverage or none at all at the price negotiation process. Consider the two following scenarios.

Scenario 1
Respondent 2 is the second lowest bidder in the short list. Respondent 2 had three responses out to RFQs. Respondent 2 has, on the day the

Table 5-5. Range of Bids

Respondent 1	$560,178.00
Respondent 2	$562,250.00
Respondent 3	$570,001.22
Respondent 4	$571,022.37
Respondent 5	$572,873.15

Network Manager called, accepted two of these three as contracts. Since Respondent 2 must hire new people to fulfill both contracts, the Network Manager should consider carefully this respondent's response to the requirement to recompute the costs of the response to the RFQ.

Scenario 2
Respondent 5 is the highest bid and has four responses out to various RFQs. Respondent 5 has a crew of seven workmen and no contracts, when the Network Manager asks Respondent 5 to recompute the costs of the response. Consider carefully Respondent 5's response to this request. Often contractors will take a contract merely to keep their employees on the payroll, taking no profit in labor and very little profit in hardware.

Another point to consider when bargaining for reduced prices is that the respondents do this every day, but very few Network Managers ever put out large RFQs more than once a year, if that. Who, then, will have heard the most often-used bargaining ploys and have more expertise in bargaining for services and material? Who makes a living by the bargaining process? The Network Manager has a disadvantage in the bargaining effort. In counterpoint, a respondent knows that he or she must shave a percent or two in most instances to get a bid, and so they build this into the price and expect to negotiate downward. If the Network Manager fails to hold out for the best price, well then, that is extra profit.

Perhaps the respondent may not want to meet the requirements for a downward modification to prices. There are other criteria to consider. Perhaps the work will be done sooner, but at no extra cost. Perhaps the respondent will substitute a higher grade of cable or connector and no extra cost. The combinations of price reductions and equipment substitutions are not infinite, but they are large and the Network Manager should explore many of them before accepting a final price.

Summary

This chapter has addressed four elements used to analyze the responses provided by respondents to the RFQ. These elements include the following:

1. The use of objective data in terms that are clear and unambiguous and understandable to any persons concerned. Subtleties must be perceived by the Network Manager and the respondent respectively.

2. The use of data analysis to provide an intermediate set of respondents (we recommend no less than three and no more than five) with whom the Network Manager must discuss other financial concerns.

3. The comparison of unit pricing of materials and cost of labor, as well as several other means of analyzing these and nonquantifiable data as well.

4. The ranking of respondents in a short list of finalists. These finalists are then asked to review their pricing in an attempt to either produce a lower price or provide better material or more services at the same price.

6
Contractual Documents and Legalities

Never promise more than what you can perform.

PUBLIUS SYRUS
Maxim 528

The RFQ and the response from the best Bidder combine to create a contract, a legal document which has specific and enforceable provisions. This chapter and Appendix C, which provides a sample contract, discusses these legal provisions. Areas covered include

- Escrow and performance bonds
- The result of a judgment
- A brief reference to the types of law which apply
- Codes and ordinances
- Transfer of ownership
- Insurance
- Subcontracting

Also certain elements of the sample contract are discussed as they are applicable to many RFQs and the resulting work. Note that from here on in the discussion, the term Bidder has been replaced by Contractor, because the status of that entity has changed.

Legal Advice Necessary

It is *very* important that the reader realize we are not licensed to practice law in any jurisdiction within the United States. What follows is a high-level review of some of the more important areas the Network Manager may wish to discuss with licensed legal practitioners. If the Network Manager's organization does not have in-house counsel, the Network Manager should consider retaining such legal practitioners to review the RFQ before it is sent out and any resulting contract which includes the RFQ by reference.

There is a close relationship between the RFQ and the Contract. Many contracts contain language such as, "The Work shall be performed in accordance with the RFQ attached hereto as Exhibit A and incorporated herein by reference." Here the RFQ changes from a purely technical, design document to a part of the contract, and as part of the contract, is enforceable under law.

The Network Manager may therefore wish to have competent legal counsel review the short list of responses to the RFQ. Often Contractors who are extremely knowledgeable in this business will add certain stock phrases to their responses as a trick, hoping the Network Manager is naive and will not understand the legal impact of this language. Many of these phrases start with the words "BMC shall..." and then go on to list things the Network Manager is contracting BMC to do in support of this contract. Another, more sleazy trick some Contractors may try is to slip in a list of definitions of terms in their response. Please note that in Appendix B, in a sample RFQ, we list the definition of terms most favorable to BMC. The Network Manager must ensure the Contractor does not attempt to change these definitions in such a manner as to be more favorable to them. A third trick that some Contractors may try is to use a phrase such as "Vendor A's Model 123ABC or equivalent." Who says what is equivalent? The Network Manager has gone to great lengths to be specific and objective in the description of the material and quality of workmanship. A slick Contractor's efforts may eliminate all of this effort. Finally, some Contractors add, as part of their response, their legal terms and conditions of employment. The Network Manager is advised in such a case to do one of two things; discard this response or ensure the Contractor knows that the Network Manager will not be bound by such terms and conditions. There are too many competitors out there who will not attempt to provide this type of legal constraint.

The Network Manager must, during this entire phase of the LAN procurement effort, keep three things in mind:

1. Who is the employer and who is the employee? The Contractor needs the work, and the Network Manager has many contractors to choose from.

2. Who suffers from a bad job, which is poorly done? In some organizations, the Network Manager may be required to submit a letter of resignation if the project does not meet the requirements or is less than useful.

3. Even if the network does not work, that does not mean BMC can refuse to pay for it.

None of these three points the Network Manager must be harsh and overbearing. Many Contractors are competent, careful professionals who take great pride in their work. But like it or not, there are incompetents in this world and sometimes they hide behind fancy letterheads or slick paper. The Network Manager should be leery of the Contractor who proclaims instant friendship based on a future contract.

In sum then, the Network Manager should have competent legal counsel evaluate the RFQ before it is sent out, review the Contractors' response from the short list, and either write or evaluate the contract which is to be signed by BMC and the Contractor.

A Contract Provides Protection to All Parties

The contract provides protection to the Network Manager and the Contractor. The Network Manager is protected by the contract's specifying exactly what is to be received, when it is to be received, and how it is supposed to work once the installation is completed. The Contractor's protection includes providing a copy of the design which is to be completed, the quality and quantity of material involved, the duration of the project, and the amount of payment to be received. Other protections include an agreement against arbitrary changes in the design without matching changes in the amount to be paid and the completion dates. There should be some mechanism in place in either the RFQ or the contract itself that addresses the way in which the design and installation can be changed. Remember, there is the unit price of the RFQ for determining the dollar values involved.

Under certain conditions a contract, at least in the theoretical sense, is a good—something that can be bought, sold, traded, discounted, inflated, or marketed. Often the contract is the basis for a construction loan from a lending institution to the Contractor. Consider the following as a possible scenario.

BMC, the organization gaining the network, is located in Progress, New Hampshire. Best Networks, Inc., the Contractor, is located in the same town. The Last National Bank and Trust (LNB&T) is the local commercial lender. BMC goes to LNB&T for a loan for its new network. Best Networks, Inc., goes to LNB&T for a construction loan for the new network it is putting in for BMC. LNB&T receives interest on the loan to BMC and Best Networks, Inc., for the same project. An interesting picture, and we often wonder how common this practice really is.

Normally the contract will contain some type of language addressing conflict resolution short of lawsuits. Both the Network Manager and the Contractor are well aware of the availability of the court system to redress grievances arising from the contract. What about mediation and arbitration? Both are useful, low-level approaches to resolving conflicts which arise from well-meaning misinterpretations of circumstances. The contract should include such processes for defusing contentious conditions.

Financial Considerations
Are Contractual

There are two legal concepts concerning the financial aspects of such a contract. These are escrow and bonding.

Escrow

Escrow is money in the form of cash or other immediately available liquid assets deposited by BMC in an account available to the Contractor under certain preset conditions. Normally the amount of escrow is the cash value of the contract BMC has signed. The money in this account flows to the Contractor if BMC fails to pay the full amount, goes bankrupt, is bought out by another firm, liquidated before the contract is completed, or another unforeseen reason. This is not a requirement that the Network Manager will run into often. Most

Contractors will not ask for escrow unless BMC has a poor financial record or is on shaky legal ground, such as in Chapter 11 bankruptcy, at the time the contract is signed.

Bonds

The Contractor may be required to provide two types of bonds: a bid bond and a performance bond. The bid bond is normally the face value of the contract. This bond is forfeit if the Contractor does not begin the work or backs out of the deal. A performance bond is money, again up to the face value of the contract, which the Contractor must provide to ensure funds sufficient to finish the job or correct any outstanding deficiencies that may arise from a poor or incomplete effort. The Network Manager should be leery of a Contractor who refuses or is unable to post either or both types of bonds. The bonds must be written by and backed by an organization whose business location is within the same legal jurisdiction as the firm (BMC in this instance) who will benefit from the bonds. The bonds must be of sufficient duration to cover the contractual period. Stated differently—if the contract states the work is to be completed by the 10th of January, the performance bond should not expire, on its face, until 30 to 45 days after this date. While many of the larger firms act as self-insurers, the Network Manager should be very leery of a self-insurer as a self-bonding organization.

There is a subset of escrow that should also be discussed. Although not the case in BMC's network, some networks require custom software development. How does the Network Manager prevent the Contractor from reselling the source and object code of this custom software? Further, how does the Network Manager gain access to this software in case the Contractor fails? The Network Manager may require the Contractor to escrow both source and object code with a third party. The Network Manager can access this code only when the Contractor is unable or unwilling to support this product. We are aware of such requirements in less than 1 percent of most RFQs. These are always very high-value products, starting in the seven figure range and going up.

Remedies for Performance Failure

Let us assume, for the sake of discussion, that the Contractor has not provided the appropriate bonding (many start-up companies cannot) but the Network Manager selected this firm because of its low bid. The

Contractor walks off the job without completing the contractual requirements. What can the Network Manager do?

First, the network must be completed. The Network Manager must select another Contractor, normally without the chance to complete the RFQ process again. The new Contractor comes in, finishes the job, gets paid, and leaves. And the new Contractor's price was $18,500 higher than the one who left. What are the next steps?

The Network Manager must, through the local court system, sue the Contractor in default. In many jurisdictions this will take years to complete. Many courts require mandatory pretrial settlement attempts through negotiation. It is difficult to negotiate with a firm that may not even be in existence. But let us suppose the negotiations are conducted without success. The defaulting Contractor simply failed to complete the job for some unknown reason. Once the case comes to court, the judge must rule in favor of one party or the other. Again, we will assume the judgment is in favor of BMC. The judgment will read something to the extent that the defaulting Contractor will pay to BMC the amount required to complete the job above and beyond the contractual price. That is, if BMC had agreed to pay $120,000 for the work to be completed, and the final bill is $138,500, the defaulting Contractor will be responsible for only $18,500: the amount over the face value of the contract.

In one instance the Network Manager is lucky, interest starts to run on the day after the judgment is made, or another date depending upon the wording of the judgment. This is about the only luck involved. Just because the Network Manager has a judgment in BMC's favor does not mean the Contractor will pay. What if the Contractor has no assets? What if the Contractor has simply disappeared? At best, the judgment gives BMC a hunting license to search through the Contractor's assets for the amount of money that can be gained through a forced sale. Assume, for example, the Contractor has a fleet of vehicles. Can BMC force the sale of these vehicles to obtain the $18,500 plus interest? Perhaps, then again, perhaps not. Some jurisdictions do not allow the forced sales of the "tools of the trade" to satisfy judgments. Let us assume the Contractor has a summer home on a lake. Can BMC force a sale of that to satisfy the judgment? Probably not, because that is a private possession, not part of the Contractor's firm.

Such lawsuits are normally brought for their nuisance value or when the plaintiff knows the defendant has large amounts of cash, or its equivalent, readily available. Most honest Contractors with this type of resource will not back out on a Contract anyway. They know it is cheaper to subcontract the work, taking a small loss, then defend a lawsuit and the nuisances that arrive with it.

Types of Law

There are three general types of law the Network Manager should be aware of in a contractual arrangement for the new network. These are the Uniform Commercial Code, state statutes, and case or common law. What follows is a very brief overview of these. The Network Manager is again cautioned to procure competent legal advice on the impact of these types of laws on the new network effort.

Uniform Commercial Code

The Uniform Commercial Code (UCC) is a series of laws recommended by committees. Many states have adopted this code as its laws concerning commercial transactions. Much of the UCC is directed toward documents, payments, financial instruments, and other fiduciary documents. Note that bonds are financial instruments, and therefore may come under the UCC in the Network Manager's jurisdiction. State statutes are laws passed by duly elected legislative bodies. These laws, although broad in scope, normally are more concerned with the rights of individual citizens in their business relationships with vendors of automobiles, insurance, real estate, or other commonly available goods and services. Business-to-business relationships may be more commonly controlled by case or common law. This is a very gray area, one where many legal practitioners are leery of forecasting results. It is much better for all concerned to have considered possible legal problems ahead of time and prepare for them with contingency plans, mandatory negotiation, and bonding.

Federal Regulations

As an aside, the FCC and federal law may have some impact on the effort if the network being purchased goes beyond the geographical boundaries of the state where BMC is situated.

National, State, and Local Codes

In addition to law, the Network Manager must worry about national, state and local codes. National codes include such requirements as the National Fire Protection Safety Code or the Occupational Health and Safety Act. By including within the contract the requirement for the Contractor to meet these requirements, the Network Manager is protecting BMC and its property from sloppy workmanship. But this inclusion, by itself, is not sufficient. State codes may expand on, counter, or ignore national codes. Again the Network Manager must include some reference to these codes in the wording of the RFQ or the contract. Local codes are the toughest of all to deal with. Often the Contractor will have to obtain a work permit to modify the wiring within the building(s) involved. This permit comes from the local codes enforcement officials, who, in turn, must inspect the final work and approve its installation. If they do not, they may have the right to immediately condemn the building(s) where the work was done, forcing everyone to find another place to work. Consider well the impact of requiring BMC to find new offices and plant within, say, five working days.

Ownership of Material

Who owns the material on site which will make up the new network when it is complete? A carefully worded contract will ensure that the Contractor owns it. Let us examine the ramifications of ownership.

The Contractor notes the requirement for seven standard 19-inch racks. These cost $223.50 each, for a total value of $1564.50. If standard procedures are followed, the Contractor owns the racks until the new network is accepted by BMC. This means if the racks are damaged, destroyed, or stolen, the Contractor is responsible for repair or replacement. If, on the other hand, BMC has accepted responsibility, BMC must make whole any damages or losses suffered. Additionally, often commercial firms' insurance will not cover such losses without expensive riders on the original policy. The Network Manager is well advised to ensure the Contractor owns all material until the entire installation is complete, tested, and accepted.

Insurance

Above and beyond the bonding required by the contract, the Network Manager should be aware of the three general types of insurance that play a part in the installation of BMCnet. These include property, workman's compensation, and professional liability. There is one additional question on insurance which will be covered at the end of this section.

Property Insurance

Property insurance is that which pays for loss through such activities as fire, theft, vandalism, or accidental breakage. The Contractor should provide BMC with a policy that will reimburse BMC for accidents or oversight by the Contractor. Assume the Contractor does work after "office hours." A tired workman leaves a door unsecured and thieves enter the worksite, stealing personal computers from users' desks. BMC's insurance may pay, but there also may be a large deductible amount. BMC could, instead of making a claim on their policy, lay a claim against the Contractor. The Contractor must then pay BMC or call in the protection in the Contractor's policy, then paying the deductible themselves. This in turn will not cause BMC's coverage to rise when that organization renews its policy.

Another form of property insurance the Contractor should provide is that which pays in case of auto or truck accidents. Assume one of the Contractor's employees drives their car to the worksite (BMC's location). Driving into the parking lot, the car slides on a patch of snow or ice, skidding sideways. It strikes a pedestrian (a BMC employee) and several BMC employee's vehicles. Can BMC sue the Contractor's employee? Of course, but what if the Contractor's employee is uninsured? Different states have different rulings on this (see state law above). Can the injured BMC employee sue BMC? The accident happened on company property during normal business hours during pursuit of company business. Can the injured BMC employee sue the Contractor? If so, does the Contractor have a right to sue BMC? Who sues who, and in what sequence? Now the Network Manager knows why it is advisable for the Contractor to have property insurance as well.

Workman's Compensation

Workman's Compensation insurance is mandated by law in most states. This is the type of insurance that pays an employee who is injured on the job, above and beyond the normal medical benefits offered to many employees. Workman's compensation is a major expense most Contractors include in determining the dollar value of the bid. In some states this is becoming an onerous load on the Contractor, and a less than scrupulous Contractor may scrimp by not having this protection. The Network Manager may wish to contact a legal practitioner to verify the following: In some jurisdictions if a Contractor's employee is injured and the Contractor does not have workman's compensation, BMC may be liable for injuries sustained by the Contractor's employee. Some courts have held that BMC would be better able to bear the loss than would the employee of a Contractor.

Professional Liability

Professional liability insurance is insurance carried by the Contractor that will pay BMC if something does not work properly. Let us assume the Network Manager designs a network which is functional. The Contractor wins the bid and installs the network as designed. However, there is a flaw in the design that allows the network to pass traffic, but at a much reduced rate. If a court holds that the Contractor should have seen the flaw and brought it to the attention of the Network Manager, then the Contractor is liable for what in the legal and medical professions is called malpractice. This very seldom comes to pass, but it is something the Network Manager may wish to consider.

Professional liability insurance is more applicable to consultants and small contractors than it is to major organizations. An independent consultant who can demonstrate professional liability insurance is showing the Network Manager that they have not only certain self-advertised skills, but another organization is willing to testify to this person's ability as a six- or seven-figure testimonial.

Insurance for Gaining Organization

Before leaving the question of insurance, there is one additional item to be considered. Many organizations are insured for all the types and forms of insurance, as noted above. If a loss is incurred, whose insurance will pay? In general it is accepted that the Contractor's insurance is to make good on the losses. But consider the following scenario.

A Contractor, in good faith, provides a $1 million face value policy for losses arising from fire, theft, or vandalism. During the installation of the new network the Contractor's employees, through carelessness, allow a fire to start, burning BMC's 10 Industrial Park Way office to the ground. The losses exceed the face value of the policy. Naturally BMC will call on its own insurance. Or should it? Should BMC sue the Contractor for the difference between what it received and what its losses were? If the losses are $1,000,100 then why bother? Legal fees for such a suit are above and beyond the $100 difference. But what if the losses are in the $10 million range? It is doubtful that the Contractor has that much money available at all. This subject was brought up for the Network Manager to consider when talking insurance with the Contractor. Is there some special device or material which is tremendously expensive and should have additional insurance? Would it be better to have it off site during the work?

Subcontracting

One additional point must be made before we conclude this chapter. Contractors will often hire subcontractors for specialized tasks. There is nothing to prohibit such actions; as a matter of fact they are quite common. The Network Manager must be aware of two things, however, and document them in the Contract and the RFQ.

First, the Contractor is specifically held responsible for the technical accuracy of the work being done by subcontractors. In this instance, the Contractor has taken upon itself the role of systems integrator and is the sole point of contact for the Network Manager and BMC in general. Often the Contractor prohibits BMC from direct contact with the subcontractors. This is the way it should be. The subcontractors are not responsible to BMC, and this must be kept clear in everyone's mind during the entire process.

Second, the Contractor must agree, in writing, that any restrictions upon the Contractor must be binding upon the subcontractor as well. This includes things such as working conditions, degree of completeness, meeting of schedules and maintaining insurance coverage.

Summary

In summary, the Network Manager must consider several items when preparing the contractual documents for the conversion from an RFQ to a contract for services.

1. First, and most importantly, the Network Manager should secure competent legal services in preparation of the documents and evaluation of the BAFO from Bidders.

2. A properly documented Contract provides protection to all parties involved by delineating responsibilities, authorities, times, dates, places, and amounts involved.

3. If the Contractor provides financial coverage for completion, it is

called a bond. If BMC is required to match this, BMC puts money into an escrow account. The implementation of an escrow account is not common unless BMC is thought to be in financial trouble.

4. Just because a court rules in favor of BMC in a lawsuit does not mean BMC will collect. A judgment is a hunting license, nothing more.

5. There are multiple laws, codes, and ordinances which govern the installation of the network and the relationship of the parties involved.

6. The Contractor should be the owner of the material on site until BMC accepts delivery of the finished product.

7. The Contractor will provide insurance for property damage, workman's compensation, and in some cases professional liability.

8. Subcontractors are to be expected in many cases. There is nothing wrong with this and the Network Manager should be aware of the relationship between contractors and subcontractors in the installation process.

7

Installation

If it don't fit, get a bigger hammer.
TRADESMAN'S AXIOM

What logically follows is the installation process. Here we will discuss the actual installation process with case studies pointing out the obvious and not so obvious errors that crop up along the way. Several of these case studies point up serious mistakes. To prevent embarrassing the parties involved, descriptions have been kept general in nature and the names of individuals, organizations, and localities have been changed to protect the ambivalent, innocent, or guilty parties. Much of this information comes from our own experience; a minor part comes from discussions with others in the industry.

This chapter follows the process of installation and discusses some of the problems that arise during

- Preinstallation
- Inspection of new material
- Staging of new material
- We will discuss problems concerning contractor's employees
- Working hours for contractors
- Installation safety hazards

- Notification of completion
- Who does testing
- Test procedures
- Test results

The reader will note the last three items in the installation are concerned with testing. This is not a mistake or exaggeration since the Network Manager must *know* the network will function perfectly when cutover takes place.

As the theme of this book has been the four-step process to network installation, we would be remiss not to mention that the installation and quality control process which is discussed here are the third and fourth steps. In this chapter the first seven case studies pertain to installation, and the last three pertain to quality control. Let us first examine preinstallation activities.

Preinstallation Activities

It cannot be stressed too strongly to the Network Manager that he or she should, just before the contract is signed, take one long, last look at the process. This final review is sometimes called a *sanity check*. A case study will be useful at this point.

Case Study 1

The firm is a medium-scale manufacturing firm with several branch offices and factories located east of the Mississippi. The corporate headquarters is located just outside Chicago. The reduction in cost of personal computer–based LANs and competition in the interstate telecommunications arena provided the firm with an opportunity to recast their network as the way it did business (which, by the way, is a mistake in itself in some situations). The corporate telecommunications manager trusted her subordinates at the branch offices and the headquarters to work out the technical details. Her background was in data processing and her experience was with the networks used to link mainframe computers to other mainframe computers and terminals. The new network was also to provide limited support to dial-up voice traffic. This installation took place before the advent of frame relay, asynchronous transfer mode (ATM), and the bridges and routers as we know them today; therefore multiplexers were required to connect the LANs to the WAN circuits in a full-mesh configuration.

The LAN managers at headquarters and at each branch office designed their LANs while the WAN specialists completed their efforts. Each branch office created a separate bill of materials in a commercially available database, and then combined their efforts. The combined bill of materials was forwarded to corporate headquarters for further consolidation and inclusion in the RFQ.

Like many other procurements, the RFQ was running behind schedule, so it was sent out without a complete evaluation by the headquarters staff. At least, that is what was reported. The organization decided to hold a bidders' conference to clear up last minute questions. It is well they did. After the introductions were made and the obligatory opening remarks stated, the first question from a vendor was, "Where are the multiplexer specifications?" A quick scan of the RFQ determined the multiplexer specifications were left out entirely.

In sum it was found that the WAN technicians assumed the LAN managers would insert the multiplexer of their choice, and the LAN managers were unsure of what kind of multiplexers handled voice well and jointly agreed to defer authority to specify the multiplexer to the WAN technicians. Unfortunately, the WAN technicians were never notified of this decision. Or, if they were, they refused to acknowledge notification.

Lesson Learned

The entire RFQ should be reviewed by someone who is knowledgeable of the technology, the process, and the desired end product. The telecommunications manager could have done several things to prevent this from happening, including the following preventive measures.

Preventive Activities

1. An outside agency can review the RFQ to provide a sanity check. This outside agency could be a friend in the telecommunications business, a consultant, or perhaps even an account executive from within the organization providing the wide area circuits necessary to link the several branch offices.

2. Although it may not be "politically astute" to do so, the telecommunications manager could have one of her most senior engineers review the final technical specifications of the RFQ. As a matter of last resort, the Telecommunications Manager can do the final quality check herself.

3. The Telecommunications Manager can appoint "Tiger Teams" at each branch office to review the work of other branch offices before the specifications are compiled into the databases. This will—beyond any doubt—create some animosity among the branch offices. But, consider which is worse: the potential for animosity or the obvious appearance of incompetence in front of many well-qualified vendors?

We have had experience with many installations both within the Department Of Defense environment and within the commercial world. In all of these network installations, if problems arose, they were at least partly based on the ego of one or more players. Because these errors were so widely spread, we have elected two case studies which clearly demonstrate the situations.

Case Study 2

In eastern central New England, a manufacturing firm needed a new network. Due to increasing product demand, a two-story building was to be completely renovated. Other buildings would follow as current office space was turned into production facilities on campus. Eventually all four buildings would be linked to a campus-wide network using fiber-optic interface repeater link (FOIRL) technology. Our effort was confined to the building being renovated.

The installation project manager, who we will call Sam, came up off the factory floor and was self-educated in telecommunications. Sam's efforts are to be applauded: Sam went through the textbooks and manuals, finding what other organizations had done in the past, but he failed to tie modern hardware into the design for the new network.

The wiring for user-to-hub/bridge locations was to be unshielded twisted pair, level-five specifications. The RJ-45 connectors were to be level-five compatible when procured. This was the first stumbling block where Sam's ego got in the way. Sam's position could be summarized as "If level-five wire exists, level-five connectors do as well." It took time and effort to prove to Sam, through correspondence with standards-setting bodies, that as of the time the network was to be installed, there were no specifications which could be used to qualify RJ-45 connectors at level-five bandwidth.

The second problem illustrated in this case study was Sam's failure to realize that there was a better choice for cross connecting than SO-66 blocks, particularly when level-five cabling was required for the transmission media. In the cover section to the budget for the project, Sam wrote down 110 type cross-connect blocks. Yet in the final bill of material, which accompanied the RFQ as the "specification," Sam replaced all 110 type cross-connect blocks with SO-66 blocks. We questioned him about this and his reply was, "Them idjits upstairs think they know my job better'n me. They can't tell me how to run my network. Put in the 66s in place of the 110s. Fools never look in the wiring closet anyway!"

Although the difference in the two types of cross-connect blocks may not affect the throughput of the data circuits attached, the 110-type block certainly is easier to change, manipulate, and, in general, troubleshoot. "Them idjits upstairs" who, by the way, were degreed engineers, did know what they wanted and where it would be placed.

We tried to explain the differences to Sam, but he was adamant that it would be "his way or the highway" for us on this project. Because of legal ramifications of termination of the contract, we tried several times to get Sam to change, but no change was forthcoming. Finally, the problem was elevated to senior management ("Them idjits upstairs") and they had to direct Sam to issue a change order to the RFQ once it was in the vendors' hands. Many of the vendors stated, at least to us, that the decision was a better one and the cost differential was small.

Lessons Learned

There are several lessons to be learned from this case study. One is that Sam's ego caused the project to be delayed twice, once for the time consumed to educate Sam where his self-education was insufficient. The second delay was when the bidders were forced to recast their bids replacing the SO-66 blocks with the 110-type blocks. Delays were not the only problem, however. Because Sam was frustrated by being overruled by senior management, actual project management was passed over to us. This was not within the contractual requirements for this project and we were caught between the horns of a dilemma. Complaining about increased responsibility to upper management would not look good for future references, yet failing to raise the problem to that level would be providing the client more than what they bargained for. No answer was forthcoming, so we took the path of least resistance.

Preventive Activities

1. There are very few preventive activities for this type of ego problem. The Network Manager must be aware that some of the people in the BMC corporation may have some ego at stake in the new network. How many of the people in BMC have some predisposition toward vendor X, Y, or Z? Each of these will try to impress their "signature" on the final effort. The Network Manager must be aware of this and prepare for ways to counter these efforts. Several additional points should be made here.

2. Workers within BMC may have already become involved through the planning and specifications process. They feel they have a stake in the process and want to keep active with the installation and testing. If the Network Manager pushes them aside, there will be hard feelings. Senior management will always want to become involved with the installation and testing. After all, it is their money that is being spent. To handle senior management, one of two directions are suggested. Most of these people have a very limited tolerance for technical details. The Network Manager can discourse, in a dry, learned fashion for hours on the impact of impedance mismatch in twists within the cable and how it can affect transmission line characteristics. If the senior management is technically literate, the Network Manager may wish to consider turning the conversation to insurance, bonds, and other business

details. The key point is to get these people out of the installation process while still stroking their egos.

Case Study 3

A large service organization in central New England was redoing its network, going from a heterogeneous, multiple vendor bridge/router/gateway mix to a more homogeneous network, using products, particularly routers, from one vendor. The organization had experienced many, many problems where a router from vendor A did not work well with a router from vendor B.

The consultant who was chosen to write the specifications for the bridges, routers, and gateways met with the project manager for the kickoff effort. After presenting observations on the problems facing the firm, the gaining firm's project manager stated, "Spec out what you want. But under no condition is hardware from XYZ corporation to be qualified for this project." The consultant was taken aback by this restriction and probed gently, trying to find out why XYZ's product line was prohibited in her client's firm. The project manager simply stated that under no condition would he even consider XYZ's products. "Anything but!" was the answer.

As the process went on, the consultant noted the project manager to be quite flexible in most areas. He was intelligent, well-educated, and in general nice to be around. It seems that he had an ego problem where XYZ was concerned. After a successful completion of the project, the consultant, the project manager, and several of his staff were having a small celebration at a local eating establishment. Seeing the project manager was content with the final product the consultant asked him again why XYZ was not allowed to be part of the bidding process. The project manager finally indicated the major reason was that his son had been fired from XYZ for what appeared to be very arbitrary reasons.

Lesson Learned

There is only one lesson here for the Network Manager: No matter how much reasoning is offered, no matter how much education is achieved, no matter how intelligent some people are, their egos still get in the way. The Network Manager must know that people who reach mid- to senior-level management within the organization can still act as unreasonably as the newest hire.

Preventive Activities

There are very few preventative procedures available to the Network Manager under these circumstances. Still the Network Manager must be aware of the situation and the problems it creates. Awareness includes the preparation to deal with the problems arising from an unreasonable ego.

Preinstallation activities give the Network Manager a chance to make very large gains with very small effort. Much leverage is available to prevent problems from arising later in the actual installation process. Failure to take the appropriate precautions during the preinstallation phase will result in many expensive failures or decisions made where the options are bad, poor, and none. As the installation goes forward, there is less and less leverage available to prevent problems and more and more effort is required to correct them. Inspection of incoming material is one of the last of the high leverage efforts the Network Manager can employ to BMC's benefit.

Inspection of New Material

Two parties inspect new materials. The first is the vendor, and the second is the gaining organization, in this case, BMC. The first case study will evaluate a problem with vendor inspection.

Case Study 4

The project involves the installation of a new LAN, and integration of this new LAN with an existing LAN. The existing LAN was wired according to AT&T's PDS wiring scheme, a wiring scheme so prevalent, it is an option as noted in EIA/TIA Specification 568. The organization gaining the new LAN wanted to ensure one-to-one compatibility with its existing wiring infrastructure, so it specified the connectors to be made up to meet the specifications of that publication.

As part of the RFQ, the gaining organization specified that the vendor would do a 100 percent inspection on all connectorized wire before it was shipped to the new location. This specifically included over 250 patch cables. The vendor created the patch cables, using his own staff and supplies. Each patch cable had a tag on it, as per RFQ specifications, noting the date and time of inspection, the inspector's identification number, and the status of the cable assembly.

The gaining organization tested samples of the patch cables and found they did not meet the requirements. When the organization checked all the patch cables, it found the cables were out of specification. What has gone wrong here?

Lessons Learned

There are two lessons to be learned here. One, and it is the core lesson, is that the vendor may hear what the gaining organization says, but is what it says what it always means? In the case above, there are two ways in which the vendor could connect the patch cables and still meet the specifications of EIA/TIA 568. The vendor chose the most common one, not the one required by the gaining organization.

The second lesson is that the vendor should have asked the gaining organization which wiring option was the correct one. The vendor assumed, in good faith, that the gaining organization would be using the more common arrangement of connector pins.

Preventive Activities

There are several preventive activities both organizations could have completed. These activities require open channels of communications and "skull sweat" on everyone's part. These are discussed as corollaries to Murphy's Law.

1. If there is an option, the wrong one will be chosen. The vendor had several ways of making sure the cabling was correct. If the vendor did not want to ask which was correct, they could have made two cables, one to each option's characteristics and asked the gaining organization which one was correct.

2. Mistakes follow ambiguity. The gaining organization should have ensured that the RFQ specifications left no room for interpretation. In this case study, we would have prevented the problem from happening by including in the RFP exact wiring diagrams showing which color-coded wire goes to which pin. This is further clarified by stating whether the connector is viewed from the front or back. A sample is shown in case study 5.

Case Study 5

The firm involved is a service organization located in northeastern Massachusetts. The firm provides services to other organizations, not individuals. As the firm had grown at more than 30 percent per year for the past several years, it needed a new building. As part of the real estate acquisition, we were assisting in the installation of the LAN which supports the firm's mission. Hardware for the LAN was selected from a well-known vendor who is also located in eastern Massachusetts. Each item which was to be part of the LAN was alleged to have been 100 percent inspected by the vendor before it left the factory.

The cable for the network was Thicknet or 10Base5 Ethernet cable—the thick yellow cable which was so prevalent for many years in the beginning of LAN installation. Each station connected to the LAN required a vampire tap, a transceiver, and a drop cable terminating in a 15 pin D connector to the user's location. The drop cables were installed in the walls of the new building as it went up. The cable was installed afterward, then tested for continuity and grounding. It passed all tests successfully. On the day of the move the computing hardware was connected to the wall outlets. The transceivers were then tied into the cable and the LAN was put to traffic.

Over 27 percent of the computers were not able to pass traffic between themselves. The bus interface units in the computers were good. These were not removed when the computers were installed in the new building. That left the drop cable and the transceivers. Based on the tests from the factory, we expected the transceivers to be good and the drop cables were suspect. This was not the case: Each computer that could not pass traffic was found to have a bad transceiver. The vendor replaced all the defective ones under warranty. The only thing which was lost was some time on the network, and because of the close proximity of the vendor, this was less than two days.

Lesson Learned

Simple documentation stating that a product meets qualifications as set out by the vendor or the RFP is not, in and of itself, sufficient proof it will work once installed. In this case the vendor admitted that it shipped a defective product but would not explain how such a defect got through its in-house quality control.

Preventive Activities

Preinstallation inspection should be considered in most cases. The Network Manager must weigh the cost of inspection versus loss of time and productivity. The key point to this lesson is the Network Manager may not be able to rely on the manufacturer for 100 percent inspection and accurate testing before the product leaves the plant.

1. Those networks which are mission essential, where it must work to ensure the safety of life and limb, or which support security functions should be inspected and tested before installation. This inspection is completed by the vendor in the presence of the Network Manager or a designated representative. The entire network must be assembled at the vendor's location and put to traffic under the worst conditions of loading or other degrading conditions. Once all the problems are worked out, the hardware is boxed and sealed before shipment. If these procedures are followed, when it is installed at the Network Manager's location, the only thing which can be wrong is the transmission media over which the traffic runs.

2. In those networks that do not meet these stringent requirements, the Network Manager may wish to inspect the products with a more relaxed approach. He could create a set of equipment as a test bed, and test each piece of equipment before it is connected into the network. This does not guarantee 100 percent operational equipment,

but it is better than betting on the inspection criteria developed by the manufacturer. The impact of the ISO 9000 series of quality-control checks may go a long way in eliminating such tedious testing.

Staging of New Material

Once the new material has been inspected and is thought to be good, it must be stored somewhere near the actual work site. Often it is stored in the building in which the work is being done. This is not recommended for several reasons. The new material can also be staged in storage trailers parked near the construction site, although this too, has problems.

The following case studies point up some of the problem areas and how to prevent BMC from suffering from them.

Case Study 6

In a manufacturing firm in west-central New England, a contractor was to install both single-mode and multimode fiber-optic cable for a LAN backbone. It was found that any metallic-based transmission media was susceptible to electrical noise generated on the floor and in some office locations.

To provide physical protection to the fiber-optic cable, the contract was to install inner duct, a corrugated plastic tube which prevented the fiber-optic cable from being squashed but still left room for additional strands to be pulled when necessary. The inner duct was brought in on 6-foot high wooden reels. The installers were to pull the duct from the reels as needed, moving the reels through the factory floor as required. Full reels were left on the loading dock; empty reels were to be broken up and put in the trash. But the production manager would not allow this to take place and required the installers to measure the run of inner duct, cut it from a reel outside the building and off the loading dock, then bring it in to install.

As can well be assumed, the installation took much longer than had been planned for, and the cost of the installation was higher. The additional cost was, quite appropriately, assigned against the production manager's department. After all, he had signed off on the original installation plan.

Lesson Learned

Not everyone who agrees with the plan understands all the ramifications of their agreement. This is particularly true with the installation process. Often cables run up an elevator shaft. This means that the elevator will be out of service for some period of time. The

Network Manager must ensure that the building supervisor or other appropriate person is aware of this service interruption. The elevator certainly must be in operation during the working day, and the installer should not use it to lift supplies from the ground floor to upper floors during the beginning and ending of the work day.

Preventive Activities

1. The Network Manager should mentally walk through the installation process, determining what the installer will have to do. If a long cable pull is required, an anchor point is needed for the pulling machinery. If concrete floors need to be cored, the installer needs power for the drill and water to cool the bit. What will running water do when the drilling is going on? Assume the installer leaves supplies and tools at the work site after business hours; who is responsible for theft and pilferage? (Hint: Read the draft RFP in Appendix B.)

2. Once the walk-through is complete, the Network Manager should invite the contractor to go through the facility and note any problems and potential dangers. The contractor should be probed regarding the quantity of material to be stored on-site, where it will be staged, and should clearly identify and secure any dangerous or pilferable items ahead of time. The Network Manager may wish to consult with local code enforcement and fire prevention officers to see what laws govern such installations and staging.

Case Study 7

In a large metropolitan area in the Midwest, a LAN installation was going on in parallel with building construction. There was sufficient storage space at the work site for a 40-foot trailer to be left there for the LAN installation crew to stage tools and materials for immediate usage. As wire and cable were used, additional supplies were relayed from warehouses within the city.

When the day was over all the tools and unused material were put into the trailer, which was then locked with a padlock. The entire site was contained within a chain-link fence and kept under guard. Pilferage was not the problem. Undetected by all concerned, sometime during the erection of the building the top of the trailer was pierced by a falling object. Water got in and the moisture corroded some of the jacks and pins on connectors and receptacles. The actual dollar value was slight, but the irritation factor was high.

Lessons Learned

Material staged outside a building is subject to loss through theft and by exposure to the elements.

Preventive Activities

The fact that the trailer was secure from theft and vandalism was good foresight on the part of the project manager. But no inspection was made from time to time to check for other types of loss, which was the responsibility of the LAN installation crew. The Network Manager should ensure that material staged outside a building is continually protected against the elements.

Merely protecting the installed equipment from damage through the elements and theft is not enough for the Network Manager and BMC. The contractor's employees, as a whole, are well-trained, intelligent, capable and willing workers. Still, there are always those one or two who will cause problems. The Network Manager must be aware of the potential problems caused by the contractor's employees.

Contractor's Employees

Of all the possible areas where problems will arise, we have found that the conduct of the contractor's employees can sometimes pose the strangest ones.

Case Study 8

A smaller service firm located in northeastern Massachusetts finally bit the bullet and made the appropriate plans to install a LAN. The building was 5 years old, 3 stories high, with about 5000 square feet per floor. A fiber backbone would tie the three floors together and extend to each of the four wiring closets on each of the floors. At these wiring closets the smart hubs would convert from FOIRL to 10BaseT for local distribution.

The existing network would continue to operate until cutover. It was a patched-up, slipshod arrangement of 10BaseT and 10Base2 technology tied together with spit and baling wire. About the best that could be said about it was that it worked, sometimes for two days in a row. The contractor offered a lower price if his employees could work during normal office hours. As there appeared to be no problems with that, the Network Manager agreed.

During the installation process one of the contractor's employees was up on a ladder in the wiring closet pulling fiber-optic cable through plenum-rated inner duct. Somehow he slipped off the ladder, dropping perhaps 3 feet to the floor. On the way down his head hit the wall with a glancing blow. It certainly was not hard enough to injure him, at least he was cursing loudly enough to prove he was alive. His (understandably) foul language raised the hackles of a senior person in the firm, and the Network Manager was called on the carpet for this contractor's employee behavior.

Lesson Learned

Accidents happen and the results of these accidents are not always what is foreseen. Normally if someone falls from a ladder, the result is injury or death. Here luck was with everyone involved.

Preventive Activities

1. Often the persons installing networks are slightly less sophisticated than the average office employee. This, along with the stress of hard physical labor, may cause the installer to misbehave to such an extent that others find it obnoxious. The Network Manager must ensure that the contractor's employees are counselled in the appropriate behavior. If not, the work should be done when the office staff is absent.

2. Parking for contractor employees is often a problem. The Network Manager must ensure the contractor knows that his employees are not to park in visitor spots, and which parking spots are reserved for senior firm members. If parking is at a premium, the contractor's employees may do better to meet at one location and carpool into the work area.

3. In some jurisdictions courts have held that a contractor is a "servant" of the firm which has hired him, and the hiring firm is thereby responsible for the contractor's actions. The Network Manager must protect his firm by ensuring two things are done. First, the RFQ must have in it a phrase which is so constructed as to ensure that the only relationship that exists is one between equals when a contractor is hired. Second, the Network Manager should in the RFQ require the contractor's liability insurance to cover not only the contractor but the contractor's employees as well. This must include insurance on privately owned vehicles that come onto corporate property.

What follows next is still another problem noted where contractor's employees create problems which are not really technical in nature, yet impede the installation of the network.

Case Study 9

We had specified and designed a large network in central New England, which spread over a multibuilding campus. The Network Manager wanted to keep us as far in the background as possible, limiting our interface with the contractor to a minimum. Once the installation actually began, the Network Manager found that issues arose that were beyond his ability to deal with on a daily basis. Unofficially we were required to take over where the Network Manager left off.

The fourth day into the installation, we were checking on the work of some of the installation crew. The schedule required that the equipment racks not be installed until the cable ladders were in place and the wiring closet walls were finished. We walked into one of the wiring closets and found two of the four racks installed out of sequence. We had to contact the job foreman, who was on another job 24 miles away and have him come in and pass on the appropriate instructions.

Lesson Learned

All parties to the contract must know who is involved with what part(s) and who has final authority. The Network Manager's ego and lack of ability to deal with rapidly changing circumstances could have compromised a large installation.

Preventive Activities

1. All parties to the process must know who has what authority, and what the limits on that authority are. This knowledge must extend to all persons who are in any way involved in the specification, design, installation, and testing steps.

2. When concurrent construction is going on with a LAN installation, the LAN contractor should meet with the construction foreman or job captain and learn the ground rules for safety. The LAN contractor also should know the job schedule so that his crew will not interrupt or cause problems with the building construction crew.

Case Study 10

A unique situation presented itself during a project where an existing 10Base2 network was being enlarged by a factor of four, and another type of LAN removed and replaced with a 10Base2 topology. The Network Manager had estimated the material needed and retained the services of a consultant for a sanity check on the plans and materials list. The consultant noted that the cable used, a generic RG58A/U was estimated very closely, leaving almost no room for error or spoilage.

When the installation was about 90 percent complete, the installation crew noted there was no cable left. The problem escalated to the Network Manager who, in turn, called in the consultant. A few calculations were run, showing the estimated lengths of cable were right and there should have been no shortage. Many times when planners estimate cable length they forget the ceiling to floor distances and wind up short that way. This was not the case in this instance. The extra cable was purchased and the project finished. The Network Manager had built in a fudge factor on the budget so that the project did not go over its forecasted limit.

Analysis showed there to be almost 800 ft of cable unaccounted for. Fortunately this is not particularly expensive. At less than $300 for a 1000-foot reel, the budget was not compromised. Where did the Network Manager and the consultant fail in their calculations? In sum, they did not.

One of the installers was taking reels with 50 to 60 feet left on them and, instead of stripping off the short remainders for the shorter runs, was taking the cable home with him. This cable's characteristic impedance is 50 ohms, which just happens to be the right impedance for CB and amateur radio antenna lines. This scrounger was making a little cash on the side with what he thought was scrap and waste.

Lesson Learned

There is a folk saying that "one man's meat is another man's poison." That is true of construction scrap and waste. Normally the people who take this type of material away see it as a noncash form of compensation and the supervisors look the other way because it eliminates some of the problems arising from removal of scrap and waste. Here, however, the installer went just a little overboard with his scrounging. That does not mean that actual theft does not occur.

Preventive Activities

These activities fall into two general categories: The discouragement of opportunists and prevention of professional thieves.

1. Opportunists are those weak-willed individuals who happen to see a pair of pliers or crimping tool laying around unobserved, or who are like the scrounger mentioned in case study 10. These people who take things know it is wrong, but they rationalize it somehow. "Ah, no problem. The boss'l get a new crimper for him anyway. Don't hurt me none." The best way to prevent this is through worker diligence and proper supervision. Tools and supplies can be accountable; the person who loses them or misuses them will replace them out of their pocket.

2. Professional thieves are concerned with large amounts of construction materials or high-value items, such as time domain reflectometers or floor-coring drills. Any high-value items must be permanently marked with the owner's identification and locked when not in use. Large amounts of construction materials should be stored in locked rooms or trailers parked at the construction site. If neither of these are acceptable, then the material should be delivered to the place of work daily; vendors deliver only that which will be used on a given day.

Contractors must work at all hours, depending upon the work schedule of the organization gaining the new network. Here is another case study discussing problems arising from restrictions to non-office hours. The hours worked by the contractor's employees are not always the normal 7 AM to 4 PM, which many technical employees follow. The installation and testing may not be completed during the same time that factory production is going on or office hours are being kept. LAN installers work nights, weekends, and holidays. This does not fit the Network Manager's schedule too well.

Working Hours

Often installation and troubleshooting cannot be completed during normal office hours.

Case Study 11

A retail grocery store was converting from previous technology to scanners with debit card readers to allow its customers to pay via electronic funds transfer. The scanner would read the bar code from the package and determine the selling price. These prices would be totalled for the customer to pay, and the items sold would be deducted from the shelf inventory. When the customer's debit card is run through the reader, the magnetic strip would be read and the correct banking transfers would take place. Naturally, the store could not shut down during the day so the new hardware was installed in a phased approach. The new cash registers, scanners, and card readers with their associated wiring would be installed between midnight and seven in the morning.

Lessons Learned

Several lessons were learned from this: One should have been foreseen, and the other was purely human oversight. The labor costs were high—that should have been foreseen. Also, contractor employees did a little moonlight shopping while their supervisor was either not looking or out of the way.

Preventive Activities

1. When installers are required to work after normal office hours the labor cost of the installation will be higher than if these people work normal hours. Whereas there is no direct guidance to be offered, the Network Manager may wish to query respondents about the surcharge involved when employees work other than normal office hours.

2. The problems with theft is not as common as many may think. Most employees are honest, and do not need to be watched to make sure they are not taking a five-finger discount. In most office and plant locations, there is little to tempt the employee with weak morals. Retail stores, on the other hand, offer many temptations. If the Network Manager is involved in work in this industry, serious consideration should be given to hiring security personnel to watch the installers.

Safety Hazards

In any network installation there are multitudes of safety hazards of which the Network Manager must be aware. These fall into two general groups: the hazards of the installation to the contractor's employees and the hazards to the workers in the Network Manager's organization. The Network Manager must be aware of the hazards to the contractor's employees, but must ameliorate the hazards to their fellow employees. In the area of safety, the federal government has a chance to stick its nose into the installation using Occupational Safety and Health Act (OSHA) regulations. The Network Manager must ensure that the installers maintain close adherence to OSHA regulations pertinent to the installation in question.

Case Study 12

A service organization in eastern Massachusetts was moving into a new building. A large LAN was being installed to support the organization in its new building. Thicknet (10Base5) repeaters were installed in the wiring closets on each floor. The wiring closets were literally closets and not much room was left for the repeaters. Instead of mounting the repeaters to the wall with strap metal brackets, the installers put shelf brackets on the narrow sides of the wall and mounted unsecured squares of plywood on these brackets. The repeaters were set on the plywood, unsecured. The installer thought that because the repeaters were above head level, they posed no danger to those who would enter the wiring closets later.

Lessons Learned

The thick cable connecting the repeaters was, as is recommended, left with large loops to ensure there was no strain on the cable or connectors. These loops were not secured to the wall, and just hung loose. The results were an accident just waiting to happen. A telephone installer was working on the 110 blocks in the wiring closet. He snagged a LAN cable and pulled the repeater off the unsecured shelf brackets. Both the plywood and the repeater fell hitting the telephone installer, ripping connectors off the cable, and bringing the LAN down in that limited area. Luckily the telephone installer had on a hard hat and suffered nothing worse than a few stitches on his shoulder.

Preventive Activities

1. The Network Manager should ensure that the plant (or office) safety staff is aware of the process of installation so that they may take the appropriate steps to make sure danger to human life is minimized. Many organizations have hazardous substances around which, to the unsophisticated installer, do not appear harmful. Yet if these substances are mistreated they can burn, explode, or otherwise injure people.

2. Tools, particularly power tools, ramsets and bang sticks can cause serious injury and death. The persons using them must be trained and subject to more than just average supervision. These tools and their associated supplies should be under positive control and away from curious fingers.

3. Depending upon requirements from the Network Manager's organization's insurance carriers, there may be a need for a dedicated safety inspector on site during the work. The Network Manager must coordinate with others on the organization's staff to ensure that all appropriate safety precautions are noted and complied with. The Network Manager who relies on the common sense of those involved is not showing common sense by doing so. Common sense is neither a sense nor is it common. Safety comes only through paying attention to details and thinking through all the results of one's actions.

Notification of Completion

This seems to be an obvious step, but very few people consider it in their plan for network installation. The job is over when it is done, right? Why should anyone be notified of it?

Case Study 13

In a massive Department Of Defense network installation, different organizations were responsible for different phases of the installation. Because the installation was so large, very formal project management techniques were applied. Company A was responsible for installing the computers and some of the computer-based communications. Company B was responsible for installing the remaining communications hardware and network software. Company C was to bring the circuits from the local and long distance carriers into the demarcation point in the buildings where the computers and networks were to be located. Naturally the local and long distance circuits would not be installed until just before they would be needed. No sense in paying for something not yet needed.

Companies A and B worked together in their installation activities. They even went so far as to loan each other labor when an important person was unavailable. Both of them finished their installations very close to the scheduled time. Their work was inspected, signed off, and they went their separate ways. Several days before the cutover was to take place, someone thought to ask about the commercial circuits. Well, to make a long story short, they had not been installed.

The work breakdown schedule noted the person who inspected and signed off on the work done by Companies A and B was also responsible for notifying Company C to begin its work. Well, it was not done, and Company C was now behind schedule (did you expect anything else from a Department of Defense activity?) by several weeks.

Preventive Activities

1. It is not only important that schedules are set, and kept, but that someone, somewhere in the organization, pay attention to the completion of milestones and major activities. This is done by requiring the contractor to notify the gaining organization of the completion of certain elements of the network. This is accomplished by making it a requirement in the RFQ.

2. Even with this requirement, sometimes the contractor overlooks it in the daily grind of installing. The Network Manager must keep tabs on the actual installation process on a daily basis. If this requires late nights and early mornings, it is all part of the pay.

Conceptually the installation of BMC's new network is complete. However, merely physically pulling the wire, installing the racks, mounting the equipment in the racks, connecting everything, and installing software is not the end of the process. Testing must take place. In the testing, BMC must consider, as a minimum, who does the testing. Should it be the installer, BMC, or an outside, third party?

Who Tests?

Once the installation is complete, the work must be tested. The sequence and content of testing is addressed in the following case studies. Here we are concerned with who tests. There are three general categories of testers: the installing organization, the gaining organization, and a disinterested third party.

Case Study 14

A medium-size manufacturing firm in southern New England reengineered its network and eliminated some of the patches and fixes which had accumulated over a period of several years. A local firm was selected to install the racks, cable trays, and wiring. The hardware vendor for the bridges, routers, hubs, and gateways would install these devices and connect them to the transmission media. After both organizations had completed their work, the new network showed several problems. Both organizations had tested their installations and found no problem.

This time the Network Manager showed more than a modicum of intelligence. She arranged a meeting of the two groups and essentially told them that they had three working days to find and fix any problems. If at the end of those three days the problem was not fixed, she would bring in an outside party to test and troubleshoot. The cost of hiring the outside party would then be deducted equally from their pay. If they wanted to get these deductions back, she would gladly meet them in court. Needless to say, the problems were magically cleared up in very, very short order.

Lessons Learned

1. The installers and vendors cannot be trusted to evaluate their work. This sounds harsh, and in reality is harsh. Human nature being what it is, people have a tendency to gloss over their own errors.

2. Testing may be performed by the installers and vendors when it is supervised or overseen by a disinterested third party or the Network Manager. By having the Network Manager evaluate the performance testing, the installers cannot allow marginal installations to slide by, leaving latent problems for the Network Manager.

3. By having the Network Manager oversee the testing, the installers or vendors may have to explain what they are doing and how this test works. This provides them with even less chance of allowing something to slide by.

4. There is another point to be considered here. If the network was designed by an outside organization or consultant, the Network Manager may think it wise to use this organization or person to evaluate the testing procedures. We think otherwise. Any design errors will show up during testing. No designers are willing to gladly admit their error. Better that design errors be found by those other than who made them.

Once it has been decided who does the testing, the Network Manager must answer the question of what is to be tested. The answer is not always as simple as it sounds.

What Is Tested?

It seems a little strange to most people to ask what is to be tested. The new installation is to be tested, right? Well maybe yes and maybe no.

Case Study 15

Earlier we had discussed a major Department of Defense installation in a case study. In this network there were over 3500 separate connections made as part of the network. To test each would have extended the installation period by a period of more than five weeks. The project manager asked the quality control staff to find a better way. They did: They identified a statistical approach to testing.

The separate connections were identified as to category, i.e., whether fiber optic, coaxial cable, RS-232 or other type. Statistically significant samples were taken from each category. If any of these failed to meet specifications, then all connections in that category were tested. This procedure saved more than four weeks of installation.

Lesson Learned

It is not necessary to test everything in the new installation if such testing would delay the completion of the project. Statistically significant samples may be tested to determine if deficiencies exist. If so, further testing is required. If not, then the Network Manager may rest assured that there is a very good chance no problems exist.

Now that the Network Manager knows what is to be tested, the next question that arises concerns the test results. How are the results recorded, what is to be done with these records, and how important are these records?

Test Results

The result of the testing is to determine whether the new network performs as was specified in the RFQ. Does the token ring network actually pass traffic at 16 Mbps? Will the routers actually route traffic at a given number of packets per second?

Case Study 16

In a large metropolitan area in the Midwest, a LAN installation was
going on in parallel with building construction. After the installation
was complete, an outside organization was brought in to test the
work. The testers had wiring diagrams that had been overlayed on
blueprints. As each separate link was tested, the testers put a red X
on the diagram, producing a blueprint covered with notes, red Xs,
and coffee stains. By the time the testing was complete, no one other
than the testing organization knew what it all meant.

Lessons Learned

1. When testing, make sure multiple copies of test forms exist. The
installing organization and the Network Manager should each get
one copy, and the testing organization also should get one copy for its
records. The copies must be clean, legible, and easy to understand.

2. The Network Manager should either design or approve the design
of the forms used for testing purposes. We provide such forms to our
clients as part of the normal design and consulting services. It is a
simple thing to prepare and prevents many problems and heartache
in the project.

3. As a given test is completed, the form is filled out and the time
stamped. This proves that on Wednesday, the 14th, at 2:50 PM such
and such a device or circuit was fully functional in keeping with
the specifications that are part of the RFQ. This provides protection
to the Network Manager, the installer, and finally to the gaining
organization. This is proof positive that the product provides the
service required.

8

Evaluation of the Effort

*That a lie which is half a truth is ever the
blackest of lies.*
ALFRED, LORD TENNYSON (1864)
"The Grandmother"

At this point in the process, the Network Manager has completed most, if not all, of the installation process. The hardware is in, the software loaded, and most of the testing completed. Now comes the time for what we call "weasel words." These include

"Yes, but...,"

"I had a brain cramp,"

"We tried the best we could," and the most infamous,

"They were outta stock so we substituted."

These four areas will then be followed by three very short sections covering

- Forms and records
- Payment and release
- A final (philosophical) review of the LAN procurement process

This chapter offers the Network Manager a preview of the results of poor installations and some of the tricks of the trade used by slipshod contractors in this business. Again, we stress these are in the minority: Very few network contractors are out to cheat and steal, and most of them are better-than-average entrepreneurs who simply have made a bad decision.

This chapter also offers the Network Manager a chance to complete the iteration process as discussed in Chapter 1. Even when the network is installed, tested, up and passing traffic, the Network Manager continues to learn what should have been done differently, what will be incorporated in the next network, and more importantly, what steps and actions are to be eliminated the next time. To paraphrase many Boston Red Sox fans, "Wait'll the next LAN!"

Excuses and Other Weasel Words

"Yes, But..."

The Network Manager has put much work into the specification and design phases of the network. These decisions are based on information received from one-on-one personal interviews with users, frontline supervisors, and senior technical and (more importantly) nontechnical staff. Additional input may have been provided by those who sell the products being installed, value added resellers and, last but not least, consultants in the industry. So what happens when all this preparation and planning is put to the test of installation?

Case Study 1

A manufacturing firm in northeastern Massachusetts was revamping their LAN infrastructure based on forecast increases in product order and output. The existing ad-hoc approach of networking was being replaced with a formal network architecture leaving room for growth. Multiple buildings would be interconnected through fiber-optic cable, both single and multimode.

Each floor of the buildings would be wired with Level 5, unshielded twisted pair (UTP) cabling. This cable would run from the users' location to wiring closets or other points of concentration. Repeaters in each wiring closet would be connected to other wiring closets using 10BaseF (FOIRL) running over multimode fiber-optic cable. The same type of fiber-optic cable would be run between buildings.

Internal requirements included the need for CATV for videoteleconferencing, distance learning, and other advanced multimedia requirements. The video signals would be transmitted over single-mode fiber-optic cable because of its inherently higher bandwidth and lower loss. The same contractor who would install the multimode cable would install the single-mode cable. The test specifications for the cable included transmission loss per kilometer and per connector. As the distance between patch panels could be measured with accuracy to the meter, it was extremely easy to determine what loss was to be expected.

What Happened

The installer pulled both the single-mode and multimode cables simultaneously. The technicians went to work putting connectors on the cables, preparing to test each run after all runs were completed. It was required that each fiber strand be tested from end to end, then connected in the patch panels. The final test would then be from one end of a given strand, a physical circuit, through all patch panels to the other end of the strand. The loss for each strand was easy to calculate, so it would be a matter of simple comparison. If the calculated loss was −3.7 dB, and the actual loss was between −4.0 and −3.0 dB, the organization would be satisfied that the installation was good.

Several of the shorter runs of fiber-optic cable, particularly those within buildings, were measured and found to be within specifications. All but two of them came in under the computed loss. One run of over 700 meters however showed a huge loss, in excess of −15 dB across all strands. To the technically sophisticated, this was an installation error; there was a kink or twist in the cable somewhere. The vendor stated the run of cable was defective from the factory. We find this difficult to believe because the same reel of cable was the source for the rest of the installation, by the same crew, using the same tools and processes. The vendor removed the "defective" cable and replaced it with a new run. The losses were still the same.

The vendor then approached the firm's Network Manager, making an end run around our technical expertise. He was able to convince the Network Manager that "Yes, the specifications require a low loss, but..." and went on from there with a litany of problems. We are surprised he did not ring in his starving children and wife. In essence he said no one could have installed the cable any better and the Network Manager should waive this specification. The Network Manager did, paying good money for a cable plant with a large, perhaps *fatally* large loss.

Brain Cramp

One of the most entertaining turns of phrase we ever encountered was from a well-meaning cable contractor operating on a shoestring budget. He cut prices to the bone to keep business coming his way and, we think, keeping his employees off unemployment. Some of his employees drove a better car than he did.

The installation this crew was working on required cables from one end of the building to go downstairs through conduit to a wiring closet. The wiring from the center of the building was to go down through the elevator shaft. The wiring from the other end of the building was to be a mirror image of the wiring at the first end. When checking the bid, we found there appeared to be an excessive amount of cable involved. We confronted this contractor with the oversight. His comment? "Gee, I musta had a brain cramp on that one!" He was able to reduce his already low bid by the amount of wire he did not have to purchase.

This example provides a light version of what could have been a damaging process once installed. The need for three different routes of cable was not arbitrary. The distances involved and the required adherence to EIA/TIA specifications did not allow the use of the elevator shaft to route all cable. What if this brain cramp had not been caught during the preinstallation review?

We Tried the Best We Could

Is sloppy workmanship acceptable? "We tried the best we could" is overworked by sloppy, slovenly, and half-baked workmen. We will not use the term *craftsmen*, or the even more generic *tradesman*.

This is a serious problem, one somewhat akin to the problem of "Yes, but...." A low bid vendor comes in and does a poor job. The specification is weak, the design is poor, the installation is sloppy, and the testing is slipshod. The result is a network which does not work well, if at all. Good material in the form of wire, cable, racks, bridges, hubs, and routers are installed and do not work. The network does not pass traffic well, if at all. Now this half-baked contractor has the gall to present a six figure bill for four figures' worth of work. What is the Network Manager to do?

There is no answer, not even a textbook answer, considering this is a book. A few guidelines are in order, guidelines which may prevent the problem from occurring. First, review the specification. Are all potential subscribers and their traffic, either current or future properly identified? Are estimates realistic? What assumptions did the person doing the specification make? Remember the leverage affect of small changes in the beginning?

Second, does the design meet or exceed the specification? Is there a need to segment networks beyond that in the specification? Does copper provide enough bandwidth, or will there be a need for fiber-optic transmission media. Is the selection of WAN circuit bandwidth correct in the price/performance tradeoff? What are the skills and abilities of the person(s) making the final sanity check of the specification and design? Really? One person may be a fine electronics engineer, but do they understand boundary routers and their application?

Third, does the equipment provided meet the equipment requirements noted in the design? We will get to the substitution problem later. Who says, beyond the vendors themselves, that a hub will pass X number of packets per second? And just how trustworthy is that source? Just because an organization has "Laboratory" or "Testing Facility" in the name does not mean it is unbiased.

Even with all this foresight and planning, the Network Manager must watch the vendor closely to ensure that workmanlike skills are used. If connections are soldered, does solder drip into other equipment? Are splices physically and electrically "clean?" Are connections tight and dressed down so that a casual passerby will not rip a plug out of a socket? Are grounds run to one location so there is a common ground to prevent 60 Hertz (cycle) hum? The Network Manager must maintain a constant overview of the ongoing work. There is no excuse for letting sloppy workmanship become part of the effort. Remember, many hundreds if not thousands of hours of preparation have gone into this effort.

They Wuz Outta Stock

In a more serious vein, we did run into the more infamous problem of "they wuz outta stock so we had'ta substitute." On whose authority? The Network Manager, BMC staff, and various other persons have spent much time and effort in the preparation of the specification, design, and RFP. Now the Network Manager will allow someone to substitute their selection for the one which had been noted? Not too likely, considering once the Network Manager accepts the final product, they are accepting the substitution. The following case study demonstrates one of the most flagrant examples of such substitution we have ever encountered.

Case Study 2

In an undisclosed organization in New England, we were requested to produce the LAN and LAN/WAN interface specifications for a service organization. The organization would have 18 agent positions, some or all to be operated 24 hours a day, seven days a week. The staffing would depend solely upon the time of day and demand for service.

The organization would query a database based on automatic number identification signals accompanying an incoming call. The response to the query would be displayed on the agent screen. The software running the query and display was to be proprietary, written in a dialect of C. The operating system was to be DOS. The only requirements for the protocol stack was that at the data link level, IEEE 802.3 is the choice, and at the network level IP is selected, and at the transport level TCP is selected.

In response to the design, one alleged vendor provided a solution with 22 agent positions, the software was Novell's NOS. When queried about the lack of compliance with the data link level, the vendor hemmed and hawed and effectively said "Don't worry, it's in there." When asked if the response met the requirements, the vendor became quite incensed, stating that Novell was a good NOS and had an excellent market share. Further, it was thought well of within the telecommunications community. All these statements are very true. Novell's product is quite acceptable. As of the writing of this book, some experts believe Novell has achieved a 70 percent market penetration. So what? The persons designing the system did not want a proprietary solution; they wanted one built on open standards. Again, the vendor substituted his judgment—we use the term loosely—for that of the customer.

Lesson Learned

If the problem had not been caught ahead of time, it is doubtful that the system would have worked as designed. One of the major problems with this particular installation was that those who were the final authority in approving it were technologically unsophisticated in the LAN and LAN/WAN areas. They knew the core business and some of the support functions, but needed to bring in several experts to cover the technological gap.

If the final authority had not heeded our direction, the system could have been installed with major substitutions and not have functioned as desired. As this was a multimillion dollar purchase, we feel external reviews saved not only this portion of the business, but many hundreds of thousands of dollars.

Preventive Activities

There are two preventive activities to discuss here.

1. Determine the need for outside experts. Does sufficient expertise exist in house? In these days of right-sizing the organization, including massive lay-offs, much expertise is going out the door. During these changes or changes to meet the market influx, there may not be time to train existing employees to cover all the areas of expertise needed.

2. Once the need for experts has been identified, pay attention to their recommendations. These people do this type of work for a living, and that is all they do. In counterpoint, the Network Manager must ensure this expert does not have any financial axe to grind and refuses to sell anything other than knowledge.

There are many, many problems which can arise based on poor or just plain sloppy workmanship by contractors. It is the Network Manager's responsibility to understand the numerous areas wherein this sloppiness can destroy the installation effort and ruin the Network Manager's reputation within BMC. Once these pitfalls have been avoided, there are other areas the Network Manager may wish to review before releasing the procurement contract.

What Documentation Is Required?

Forms and Records

Once the network has been tested and accepted, the forms and records that are part of the testing process become part of the warranty. The Network Manager should have negotiated a warranty or guarantee period on the components which make up the new network. If the vendor argues that equipment is deficient from the factory in some way, then the Network Manager has definitive proof that such is not the case.

These forms and records become a baseline for future network management functions. Often the tests take place just after the network is put to traffic as a real world exercise in determining network capacity. As such, the Network Manager will learn that WAN circuit 74FDDA78901 will not pass traffic at 56 Kbps as it is specified to do, but only at 53.5 Kbps instead. This type of information will be a guide for the Network Manager or a successor when additional demands are placed on the network in the future.

Finally, these forms and records can become evidence in civil suits for contractual misperformance or nonperformance. Many times when the vendor realizes that the product does not work as specified and they must replace it at a large loss, they may stonewall, bluff, or otherwise try to evade responsibility. Tests which show a router has a throughput of 6000 packets per second when the manufacturer boasts of 12,000 packets per second is quite convincing to even the most technically unsophisticated layman.

Payment and Release

The method of payment for most installations is a lump sum upon completion of acceptance testing. Often small systems integrators need payroll money and money for suppliers as the work is being done. Yet if the Network Manager pays out good money before the work is done, and the work is done wrong, there is small chance of regaining it once it is paid out.

The usual procedure for partial payments goes something along these lines. At the first of three prenegotiated milestones or dates, the Network Manager pays over 30 percent of the total amount due under the terms of the contract. At the second set of milestones or dates an additional 30 percent is paid. The third prenegotiation point is normally that of the end of the testing procedure. This leaves 10 percent outstanding.

The final 10 percent is paid normally 30 days after acceptance of the network. This gives the vendor motivation to ensure any discrepancies are corrected rapidly and to the best of their ability. This 10 percent normally is the vendor's profit for most installations. Many vendors are aware of this 30-30-30-10 spread and create their bids accordingly. One vendor, in a moment of alcoholic indiscretion, let it slip that if he was cut out of the 10 percent final payment, he would still make a small profit.

One additional point to cover having to do with terms of payment is that most judicial authorities hold that if the Network Manager has paid for a network, then the network is acceptable and the burden of proof otherwise falls on the Network Manager, not the vendor. Again, the Network Manager is strongly encouraged to seek professional advice from an attorney licensed to practice in that jurisdiction. In sum, payment is thought to be a release for all responsibilities not contractually specified in terms of warranty or guarantee.

Summary

There are wheels within wheels, circles within circles, and stories without end. The selection, design, procurement, and evaluation of the LAN within most organizations fits into any of these categories, particularly within BMC.

In an organization the size of BMC, the time from LAN inception to a LAN ready to pass traffic can be no less than 90 days to be done well. We would be more inclined to think somewhere in the 120 to 150 day range. The question any network manager or MIS manager must ask is how much will the organization itself change during this 90 to 150 day period. Will the LAN, which was specified in November, be large enough and flexible enough to carry the traffic that is offered in April? In counterpoint, is the LAN going to be too large and complex for April's organization?

The Network Manager must remember that the LAN procurement process is iterative and is never finished, merely reaching a plateau of functionality based solely upon the match of the LAN's characteristics and the organization's needs. The other side of the plateau becomes too visible, too soon.

9

Project Management

I have come to put you back on schedule.
DARTH VADER
The Empire Strikes Back

The Network Manager should realize that small projects, at the back of the envelope level, do not need much of the information provided here. In counterpoint, larger procurements require formal project management with all the problems associated with such efforts. This chapter discusses the following elements of project management which apply to smaller network installations. These include:

- The management triangle of time, money, and people
- The critical path method and how to compute it, including derivatives such as Gantt charts and milestones
- Classes of issues and their workoff processes
- Number and type of required reports

The Project Management Triangle

In any project, there are three elements which define and constrain the project: time, money, and labor. This section addresses the relationship

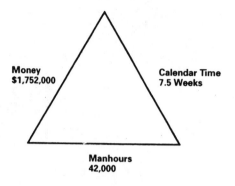

Figure 9-1. The management triangle.

between them and how changing one must change at least one of the other two legs of the triangle.

Figure 9-1 shows the three elements necessary to install a theoretical medium-size LAN in a medium-size organization. There are 35 people working 40 hours a week for 30 working days to obtain the 42,000 man-hour figure. The actual calendar week passage is close to 7.5 weeks depending upon holidays, work stoppages, and other uncontrollable incidents. The dollar figure comes from information developed in the specification and design phases. It includes all expenses for, among other things, parts, labor, and transportation charges. These figures describe the project the day the contractor arrives to begin work. Let us see what happens when circumstances change.

Scenario 1. One week into the project, one of the following incidents occurs: a blizzard, hurricane, tornado, fire, or earthquake. The incident does not destroy the work site, but work is delayed by three days while the roads are cleared and repaired. This has moved the project completion date three days past the scheduled completion date. Let us examine some of the consequences of this action.

The most obvious consequence is that the LAN will not be ready for operation on the day it is forecasted. If this construction is part of preparation for occupying a new building, the users may not be able to occupy the building. They must stay in their current building at the increased cost of renting or leasing. An interesting problem arises concerning salaries. For three days the entire work force is willing but unable to work. Do they get paid, even if they do not show up on the job? Will the project funding allow for this? Assume the average salary for these people is $10 per hour. Multiply 35 (number of people) times eight (work hours per day) times $10 (hourly rate) and the total project budget just increased by $2800. Does the project manager ask the contractor to eat this cost? Let us assume the workers do not get paid for days not worked. Let us further assume that no one will allow the end date of the project to slip. The lost time will be made up in overtime. Overtime is normally 1.5 times regular time. Therefore the budget increases by $2800 multiplied by 1.5 or a total increase of $4200. Notice however that whereas the total number of worker-hours has not changed, the time leg has changed from 30 days to 33 days and the money leg has changed from $1.752 million to $1.756 million (worst case).

This is the key point to remember when dealing with the project management triangle. Moving one leg *will* affect the length of at least one other leg to compensate for this move.

Scenario 2. The contractor notes an overage in their bid, and, being honest, tells the project manager that $22,000 will be "returned." What will happen to the extra money? The obvious disposition of this money would be to reduce the money leg by that amount. This may not be a good idea. Subtracting the $22,000 from the $1.752 million money leg leaves $1.730 million as a total amount. By spending this $22,000 for extra workers, the installation time (the time leg) could be reduced accordingly.

The previous two scenarios discussed what happens when the changes are affecting the time and money legs of the triangle. The man-hour leg and its impact on the project management effort is the final scenario of this section.

Scenario 3. Due to a miscalculation in the specification and design phase, the project manager determines that additional worker-hours are required to complete the project. There are two ways in which additional man-hours can be procured: increase the duration (the time leg) of the project or hire additional people to meet the schedule. There is no one right answer to this situation. If the project goes beyond the scheduled completion date, the network may not be ready. The impact of that was shown in Scenario 1 above. This not only delays the completion date, it will increase the cost for labor by some amount.

Hiring additional people will cost more but will allow the project to be completed on the time it is scheduled. Hiring additional people is normally more expensive than using the same number of people for longer periods of time. This is a rule of thumb and is subject to change at any time. Despite which approach is selected, at least one of the two remaining legs must be modified.

Summary of this section is quite succinct. Change any one leg of the project management triangle and at least one of the remaining two legs also must be changed to reflect the results.

Critical Path Method

The critical path method is a legacy of the birth of formal project management in the World War II era. When creating an estimate for the cost and duration of the project, certain mathematical calculations are necessary. These calculations then provide the Network Manager with milestones and other key checkpoints along the way. There are relationships among elements of the network installation project which also will be explained. Here we will discuss the

- Beginnings of the project
- Work breakdown structure
- Overall project management
- Graphical representation of the effort
- Cross unit relationships
- Final step in the unit listings
- Milestones

Beginnings

Any project starts with a goal in mind. In this instance the goal is a new LAN for the gaining organization. We know that the project will consist of the four steps discussed previously: specify, design, install, and test. For the sake of simplicity we will begin our calculation with the specification process.

Work Breakdown Structure

Any job can be broken down into unit levels, e.g., an action that is inclusive unto itself. A unit can be something like putting a connector on a cable or developing an interview form. What are the generalized work units involved in the specification phase? Development of the survey or interview form to be used, gathering the necessary data by interview, analysis of the data to determine the results, and writing the formal specification are four of them.

We will use a survey or interview form as an example. The steps to developing such a form are:

1. Determine what knowledge is needed.
2. Phrase questions to obtain that knowledge from interviewees.
3. Review these questions with others for their evaluation.
4. Write the actual interview form.
5. Edit the final form for grammar, spelling, and punctuation errors.
6. Duplicate the form.

Keeping in mind the final effort is to determine duration and costs in preparing the interview form, the following mathematical functions are to be performed. Obtain three estimates of the duration of each of these tasks from those who are knowledgeable in this area. We will take the first task, "Determine what knowledge is needed" as the first step. The three estimates are shown in Table 9-1 in ascending order. This is the way they must be shown.

Multiply the middle estimate, Estimate 2, by a factor of four. This provides a total of 12. To this, add Estimates 1 and 3. This provides 2 + 12 + 3.5, which equals 17.5. Divide this by 6 for a result of 2.92 days. Now we know the duration of the unit of "Determine..." will be 2.92 days. Table 9-2 shows these steps applied to the unit of developing the interview form.

Table 9-1. The Three Estimates

EST 1	EST 2	EST 3
2 days	3 days	3.5 days

Table 9-2. Estimated Time for the Survey Form Completion

Step	EST 1	EST 2	EST 3	Duration
1. Determine...	2	3	3.5	2.92
2. Phrase...	1	1.2	3	1.6
3. Review...	1	2.1	4	2.23
4. Write...	2	3.3	3.5	3.12
5. Edit...	0.5	0.75	1	0.75
6. Duplicate...	1	1.5	1.75	1.46

Adding the Duration column for all steps of this effort indicates that the best estimate for this unit is 12.08 days. This should be rounded down to 12 days, or a total of 96 man-hours of effort. This notation as man-hours becomes important as the process of estimating goes on. Remember, although man-hours and calendar days are both measurements of time, they are different in the critical path function.

There is an important question the project manager must ask at this point. Of the steps listed here, which one(s) can take place in parallel? Of all these, only the Review and Write steps can take place in parallel, and then only when the number of questions is very large. If, in the project noted in Table 9-2, the Review and Write steps are done in parallel, the duration of this item of work is 9.85 days, rounded up to a total of 10 days.

Project Estimate

The project estimate is the total of all WBS unit estimates. It is key to remember that this is not a single, serial process, but many serial and parallel processes. Let us examine some, but not all, of the parallel processes which take place with network procurement.

Parallel Processes During the Specification Step

1. Determine numbers and types of people to survey.
2. Prepare survey instrument.
3. Arrange travel for those performing the survey.
4. Create the project management plan.

Parallel Processes During the Design Step

1. Determine equipment placement within the buildings.
2. Prepare bill of material.
3. Create test procedures.
4. Begin the procurement document.
5. Obtain licenses and permits.

Parallel Processes During the Installation Step

1. Site all material at the desired locations.
2. Begin rack/closet/cabinet installation.

3. Run all transmission media, leaving large loops on the ends.
4. Inventory all hardware/firmware/software as it is delivered and installed.

Parallel Processes During the Testing Step

1. Create database of physical, virtual, and network addresses.
2. Identify or mark hardware and transmission media as it is tested.
3. Connect users to elements of the network which have been approved.

This is not an all-inclusive list of parallel activities. It is a list of activities which applies to many large network installation projects and even at that is very general in nature.

Next Step

What does the Network Manager do with this information once it is compiled? Let us go back to the list of six steps noted in Table 9-2 above. Instead of putting them in a tabular format, let us put them in a graphical format in Fig. 9-2. Down the left-hand side we see the six items to be complete. Across the top (bottom) of the graphic we see the calendar dates.

Note that in Table 9-2, we found this process required 96 man-hours. Therefore the duration noted should be 96 man-hours over the calendar dates shown. The graphic shows only 80 man-hours are being used. Why? Look at the two items under Write and Edit. These two can be performed in parallel, thereby reducing the total *duration* but not number of work-hours of this unit.

Predecessor/Successor

The various units of work within the specification phase are not all in parallel. Once the survey has been completed, it is administered. Here

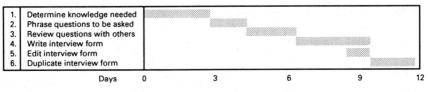

1.	Determine knowledge needed
2.	Phrase questions to be asked
3.	Review questions with others
4.	Write interview form
5.	Edit interview form
6.	Duplicate interview form

Days   0   3   6   9   12

Figure 9-2. Estimate of time.

are two important definitions: predecessor and successor. The survey preparation unit is a predecessor to the survey administration unit. A successor cannot begin until a predecessor is complete. At a much higher level, the design activity is a successor to the predecessor activity—specification. But the design activity is a predecessor activity to the successor activity of installation. Therefore a particular action can be both a successor and predecessor, depending upon its relationship *in time* to other actions.

Cross Unit Relationships

Certain predecessor/successor relationships may extend across work units. A preliminary design review (PDR) may begin before all the survey information is complete. The PDR may address only the connectivity needed, not the bandwidth and performance required.

Final Step

Once all the work units have been identified and the predecessor and successor relationships noted, they must be summed graphically. The final representation will have a list of all work units down the left-hand side, with calendar dates extending across the top (or bottom) of the sheet. Because most projects large enough to require formal project management handling are so complex as to be extremely difficult to manage by one person, software has been developed to control the planning and implementation process. The description and explanation of what this project management software actually does is beyond the scope of this book. The Network Manager is strongly encouraged to seek out vendor literature and product reviews in the selection of project management software.

Milestones

The milestones are those work units that have been identified as important. The milestone is the end of a work unit. A typical milestone may be the end of the survey preparation process or the end of the specification step. There are no hard and firm rules for the creation or identification of milestones. The key point is that the contractor(s) must be aware of them and the reports which surround them.

Review

The estimation of the duration of the project is a purely mathematical process, depending upon the knowledge of experts. These various estimates are linked horizontally in time, in what is sometimes known as a Gantt chart. The end of some of the work units is a milestone.

Issues

Issues will arise beginning with the release of the RFP and will not end until after the installation and final test of the new network is completed. Despite the time an issue is raised, its treatment is similar across the duration of the project. This section will introduce the concept of levels of issues and their separate treatment. Issue review team makeup will also be discussed, and issue documentation will close this section.

Levels

Issues normally can be categorized into three levels. For the purposes of this publication these will be identified as levels A, B, and C.

1. A-level issues are those which if not addressed immediately will terminate the project short of completion. These issues normally concern items such as time, labor, patent/copyright infringement, right of way, or other political or legal concerns. The key identifier of level A issues is that they will not go away by throwing money at them.

2. B-level issues are those that are technical and operational in nature. These include areas such as protocols, topographies, construction details, and to a lesser extent, money. Almost all level B issues can be resolved through application of money in the right places. Safety is normally a B-level issue, but may, under certain legal standards, become an A-level issue.

3. C-level issues have no impact on the time of completion and the technical quality of the final product. Normally these are procedural or arise through misperceptions by those not fully cognizant of all elements of the plan.

Issue Work-Off Procedures

Although the actual makeup of review teams is discussed below, these teams' actions are important to the issue level process and procedures.

By identifying issues in three categories, this action sets the priority of work-off. In theory, all A issues will be worked off, then all B issues, finishing with all C issues. In reality often level A issues are held in abeyance waiting the response from lawyers, accountants, or very senior management. This gives time to attack other levels of issues while awaiting decisions on A-level issues. What follows are the steps necessary for working off any issue. For the purpose of this explanation, we have selected a theoretical issue from the B level: Throughput on routers.

Issue.   Noncompliance with specifications for throughput on routers.

The network RFP specifies throughput on any port of all routers at 12,500 packets per second, assuming 1518 bit packets. The vendor's product meets the specifications—on paper. When tested, the throughput is 11,274 packets per second using 1518 bit packets. The problem is reported to be because the vendor's description fails to account for latency of traffic through the router.

Action.   Apply the four-step problem-solving process.

1. Determining the problem. Is the problem actually the vendor's misrepresentation, or is there a problem with the testing procedure? First, we recommend getting a copy of the vendor's testing process or discussing the testing process with them on the telephone. Perhaps the testing organization did not set parameters correctly or had an earlier version of software loaded. Perhaps the equipment being used to test the router was in itself defective and needed to be brought up to full operating potential.
2. Identify all possible solutions. Once the problem is determined to actually be a router which is not up to specifications, the review team must list all potential solutions. The first two solutions which come to mind are to create a study committee to evaluate the impact of this shortfall, or scrap the entire process and redesign the network. These are two ends of the possible spectrum of choices and as such, selection of them should be viewed with much skepticism. Some possible solutions include (in no specific order):

   a. Replace this vendor's product with an equivalent product meeting the same specifications for about the same price. In most instances, the contractor will absorb the minor losses from this.

b. Ignore this slight discrepancy—the most likely solution.

c. Install one or more routers in parallel to provide greater throughput. Certain engineering problems may arise from this.

d. Subnet and/or segment the network further to allow for greater throughput. Certain operational and network management problems may arise from this.

3. After the potential solutions have been identified, they should be rank-ordered based on several factors: including cost, how quickly the problem can be solved, integration into the final network, and the impact on the corporate culture. Assume for instance that replacement of the router by one from another vendor was selected. The replacing vendor does not support its products in the field. This means that the Network Manager must train staff members to support this device. Does the Network Manager's organization's corporate culture support such a decision?

4. Implement the decision. Although this appears to be a simple step, it is fraught with danger. If the issue review team is not technically sophisticated or is controlled by a person with limited technical depth, then the solution implemented may create more problems than it solves. By rank-ordering all possible solutions, the availability and implementation of a secondary or tertiary solution does not require additional effort by the review team.

Issue Work-Off Completion

Like any other formal management process, the issue work-off is not complete until the paperwork is done. At a minimum, the issue work off documentation should list

1. The issue itself

2. How the issue was raised

3. Solutions identified

4. Solution(s) applied

5. Date of closure of the issue

To ensure accountability is established and maintained, the issue review team leader must sign the document indicating the issue is successfully closed. We have had more success when all members of the issue review team are required to initial or sign off on their work.

Issue Review Team

As there are several levels of issues, there should be several different types of people on the review teams.

Content of A-level issue review teams. This type of team may be made up of those persons who are

1. Knowledgeable of the organization receiving the network
2. Knowledgeable of the industries involved,
3. Knowledgeable of management theory and practice
4. Currently practicing law
5. Experienced with network installation practices and procedures

Each of these people and their impact are discussed below.

Those Knowledgeable of the Organization Receiving the Network. These are senior people in the organization at normally vice presidential level and above. They are not technically qualified, except by accident. These people are concerned with market share, profit margin and comparison with competition in the market place.

Those Knowledgeable of the Industries Involved. The industries involved are not the competition, but the suppliers of products and services which are used in the new network. Two general types of people fit this description: consultants and those who are long-term sales types and hardware or software application specialists.

Those Knowledgeable of Management Theory and Practice. The results of issue work-off actions will have an impact on the way in which the organization gaining the network performs daily. Often these decisions will have a major impact on the way in which the organization is managed. Consider the impact of videoconferencing on interviewing job applicants. Again, we consider that lecturers at colleges or universities or perhaps consultants would be the best type of person to become part of the A-level issue work-off teams.

Those Currently Practicing Law. An attorney may or may not be required. This is a decision made at senior management level, and as such is beyond the scope of this book.

Those Experienced with Network Installation Practices and Procedures. There are many tricks of the trade which contractors

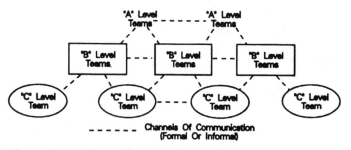

Figure 9-3. Bridging of teams.

who have much experience with installations, or equally experienced consultants, can bring to the issue resolution process. But because this is knowledge developed over many years of hard, bitter experience, these people will charge serious fees for such advice and guidance. Still, at times there is no other source of direction, and the Network Manager must be prepared to spend money for such guidance.

One or more of the people who are on the level A review team, and who understand technology, also should be part of the B-level issue review team. This will provide continuity of solutions and open channels of communications between teams. This bridging of teams is duplicated between the B- and C-level issue review team. This is shown graphically in Fig. 9-3.

Content of B-Level Issue Review Teams. Team members at this level should be knowledgeable in the areas of

1. Network hardware

2. Network software

3. Building construction trades

4. Technical network management

5. Project management

In general these team members are not management specialists, they are more concerned with the bits and bytes aspect of networking. They may know what a budget is but have no experience in creating one. Many of them think turnover is a type of pastry and not a personnel management metric.

Network Hardware. Network hardware knowledge includes the capabilities and limitations of devices such as bridges, routers, servers, multiplexers, or other ancillary devices. This also includes things such as throughput, reliability, and general operational problems. This person is probably a degreed engineer, very senior technician or experienced network consultant.

Network Software. The person who will have this responsibility is knowledgeable in network operating systems, firmware in some devices, what is euphemistically known as *middleware*, network applications such as e-mail or e-mail–enabled applications and should have some experience as a systems administrator. This person may not be a degreed engineer but must have quite a bit of experience and perhaps be certified by some vendor or independent agency.

Building Construction. This type of knowledge does not include formal education in architecture, civil engineering, or any of the trades. Rather, this person should come from the custodial elements of the organization, particularly someone involved with low-level repairs to the electrical, HVAC systems in the building(s), or plumbing and carpentry. This person will have knowledge that derives from experience in the organization gaining the network which can be used to determine the placing of devices and the running of transmission media.

Technical Network Management. This person will be the supervisor of the help desk in a large organization or the network manager in a smaller organization. This is not a position for the Chief Information Officer or MIS Manager in any but the smallest organization. A well-seasoned consultant in the area of network management may be considered, but we do not recommend it. One of the most important criteria in selecting this person is the knowledge the selected person has of the organization receiving the new network. Most independent consultants do not have such knowledge.

Project Management. There are many writers and theorists in the subject of project management. There are very few who have the extensive hands on knowledge necessary to act as a member of an issue review team. Certain management consultants who are technically oriented may be able to fill this position. In most cases we recommend this position be filled by the Network Manager. Not because of that person's inherent knowledge of project management, but because of their detailed knowledge of the project and the variables involved.

Content of C-Level Issue Review Teams. Team members at this level should be knowledgeable in the areas of

- Ergonomics
- Appearances
- Grammar

People handling these issues are not telecommunications technical specialists, but are knowledgeable in other areas.

Ergonomics. This is the man-machine interface at a very low level of importance. We are not talking about keyboard design or reference lines on the front panel of the device. Issues in this area include sufficient lighting for operation and maintenance, spacing of racks so that technicians can get around them, or sizing of wiring closets. Probably the best person for this position is a consultant in the area of industrial engineering or safety and hygiene.

Appearances. Often the equipment, including servers, bridges, and routers, is concealed in closets out of public view. Equipment racks and patch panels are scattered haphazardly around the area with the understanding that only the technical support staff will ever see them. Occasionally, some organizations use such installations as glitz and glamour for VIPs. Issues will arise concerning the appearance of such installations. The persons who should be working on issue resolution are the ones who require such a display of technology. They should be assisted by one or more persons who are technically skilled in network design. We have found that those who are concerned with appearances do not understand the rules of design which mandate equipment locations within the boundaries of the network.

Grammar. As unbelievable as it may sound, we have seen people at the vice-presidential level being forced to clear up issues based on grammar in documents. We do not understand why a dangling participle should take up five minutes worth of time of a person who is making $90,000 a year for managing the entire production effort of a multi-million dollar corporation. Perhaps grammarians are not needed to solve such issues, but merely someone who has some grasp of sentence construction and a forceful personality.

Documentation. Several types of documents are necessary for the proper control of issues. One of these was discussed previously when discussing the implementation of the solution. Here we will lay out the various forms and records required for issue control.

The Issue Document. This is a statement that fully identifies the issue being raised, its actual and potential impact on the project, and recommendation for classification. Any person involved in the project should have the right to raise an issue. The project manager will determine whether the issue in fact exists, what its level should be, and what procedures should be followed in handling it.

The issue will be fully documented. The source document, e.g., the RFP, contract, or other formal paperwork will be identified by title, date, page, section, subsection, and paragraph number. The name, address, telephone, and facsimile number of the person raising the issue should be included. If e-mail, LAN, or Internet addresses are available, they should also be included.

Hardware, firmware, and software that are part of the issue will be identified by make, model, serial number, revision, or other unique identifiers. The location of small items such as EEPROMs or diskettes may have to be identified, e.g., file cabinet in third-floor wiring closet.

How the Issue Was Raised. Issues arise from mistakes. They could be mistakes in design, in writing, in perception, or unforeseen changes. Those who learn from their mistakes are better in the end. Therefore reasons why the issue arose must be recorded. Was the mistake found in the preliminary design review? Did the issue arise during specification, design, installation, or testing? Who found the mistake—a contractor, an outside party, someone from the organization, or a vendor? How was the mistake found—through review, study of the project, an accident or personal injury?

Solutions Identified. This can be a lengthy document. It will list all the potential solutions which were brought forward and the discussion surrounding them. Often there will be such supporting documents as cost estimates or labor estimates attached. The content is normally in that of minutes of a meeting where each speaker is identified and their comments are noted, not verbatim, but in substance. Counter arguments are included and where applicable, voting is recorded.

Solution(s) Applied. The sole purpose of identifying possible solutions is to eliminate the problem, thereby eliminating the issue. The Network Manager may wish to pay particular attention to the process used to apply the solution. There is a difference between selection of a solution and application of the selection. Just because the best solution is identified does not mean that it will be correctly applied. If the solution is to purchase a replacement product and the purchase order never gets out of some manager's in-box, then the solution has not been applied. This may seem elementary to many people, but things do get overlooked occasionally.

Date of Closure. An issue is closed when the issue review team says it is by agreement. This agreement may be oral, but must also be in writing. There is no requirement for unanimity amongst the team, but it does help. If unanimity on the fact that the issue is closed is not reached, the dissenters should provide a written dissent noting specific reasons why they feel the issue is still open.

Review

Issues are broken down into three separate levels. More can be created, but we have seen that more than five causes additional complications. Three levels seem to be sufficient. Issue review teams are necessary; the skills of each team will depend upon the level of issues being worked. Finally, the documentation necessary for issue control was discussed.

Reports and Project Management

Previous sections of this chapter have covered, in some depth, the need for determining when things should take place, and in what sequence these things should happen. This section discusses the way in which the Network Manager becomes aware of what has and has not happened, and when this has taken place. Unless otherwise noted, all reports should flow to the Network Manager or that person's designated representative.

Each of the following subsections describes the title and content of the report, who creates it, and when it should be submitted. These subsections are in alphabetical order.

Costs

Any project has some amount of funds attached to it. The contractor must report to the Network Manager material, salaries and other miscellaneous items. This report can be part of periodic reports discussed below, or separate, depending upon the duration of the project and magnitude of costs. The contractor and the Network Manager should hash out the details for such reports early in the contract negotiation process. The format for such reports is normally a listing of budget codes and expenses charged against those codes on a cumulative basis. Normally cost reports should be provided on a biweekly or monthly basis.

Delay

The delay report is a textual notification that one or more milestones will not be completed as scheduled. It is created by the contractor, most probably by the job captain or foreman, and forwarded up through the contractor's channels for release to the Network Manager. It must identify all milestone(s) which will be affected, how the delay will impact the project overall, the actions taken to abate this impact, the cause of the delay, corrective actions taken, and finally, what steps will be taken to prevent this from occurring in the future. Approximate dollar value impacts also will be provided where it is possible to compute them. This report should be submitted after the delay is confirmed, but before all possible corrective actions have been taken. The Network Manager may be able to help the contractor overcome the delay. Many contractors, however, do not like to provide these reports as they feel such reports show incompetence on their part.

Legal Actions

The contractor will provide the Network Manager with written notification of all actions filed against him within the area of jurisdiction in which the Network Manager's organization is located. The purpose of such a report is to ensure the Network Manager knows of such pending actions and what their impact on the project could be. Often this type of information is sensitive, but it is not proprietary; contractors cannot hide behind the shield of private information. Most legal filings are public documents and, as such, cannot be hidden from public view. Usually such filings are not publicized unless there are ulterior motives for such actions. The contractor should provide this report in writing, within 24 hours of notice of filing or legal service and notification, whichever comes first.

Lost Time Injuries

The lost time injury report content and format may be controlled by OSHA and other state, county, and possibly city requirements. The contractor's insurance provider(s) also will need to be informed of this incident. The Network Manager should be informed through the mechanism of copies of documents provided to other organizations and entities. The Network Manager or designated representative should be informed by telephone of the incident as soon as practicable after treatment has been rendered to the injured person(s).

Material Deliveries

The contractor will provide, as part of the project management schedule, the forecasted delivery dates of material to either a storage facility or the construction site. Failure to deliver this material will, at a minimum, set the schedule back by the number of days of nonreceipt of the material. With this in mind, the Network Manager must demand and expect reports within 24 hours of the scheduled receipt of material. Here the Network Manager may substitute an exception report; e.g., the only reports rendered will be those which indicate material is *not* delivered on time, in the correct quantity. These error reports will be textual in nature, probably enclosing bills of lading, delivery receipts, or other forms and records attesting to identification and quantity of material.

Milestones

Certain milestones are identified in the project management plan that the contractor should provide. The contractor must provide a report indicating the (non) achievement of these milestones. These reports should be provided within one working day (plus or minus) of scheduled completion. The form and content should be decided upon during the initial working meetings between the Network Manager and the contractor. At a minimum, these reports will be textual and may include charts, graphs, and pictures as needed.

Periodic

A periodic report normally is a weekly report, summarized monthly. This is prepared by the contractor, listing all incidents, accidents or other noteworthy subjects which have occurred since the last report of this type. Often the purpose of this report is nothing more than a paperwork drill to ensure the contractor is doing what the contract requires. Still, such reports can also be a good management tool, making both the contractor and the Network Manager aware of shortcomings early in the project. These reports also provide a paper trail which will support both the contractor and the Network Manager in case of lawsuits or negotiations concerning misinterpretations of contractual requirements. Periodic reports also can be a tool for conveying knowledge of financial expenditures or encumbrances of funds.

Phase Completion

Major network installation projects have phases as well as milestones. A phase may include a specific action by a subcontractor; it may depend upon weather-related activities. For example, it is difficult to bury cable in frozen ground. A phase also may be the completion of effort in one geographic area. Phase completion reports can be separated from milestone reports by a very simple analysis: Phase completion reports are normally accompanied by bills. These reports and bills are created by the contractor according to criteria which should be agreed to during the contract negotiation activities.

Property Damage

Almost every project will have some type of property damage accompanying it. This may be as inconsequential as a sloppy worker knocking a coffee cup off a desk and breaking the cup to as serious an event as a fire, which levels the building in which the new network is to be installed. The contractor must notify the Network Manager as soon as practicable after the immediate corrective action has been taken. Often the contractor's insurance carrier requires certain forms be filled out. The contractor may wish to attach these forms to the textual narrative describing the incident. Photographs of damaged property and goods also should accompany any reports where the dollar value of losses is high. Property damage reports also should be included for damage to property belonging to the contractor. This is particularly important when the property which is damaged is actual network material such as bridges, routers, or transmission media.

Shrinkage

The Network Manager must be notified by the contractor of any loss when the loss is known. The cause of the loss does not have to be identified immediately. The contractor and the Network Manager must work together closely to determine what was lost, when it was lost, and possibly who was responsible. If the loss was through theft, the need for confidentiality should be self evident. The loss should be communicated by telephone to the Network Manager, then followed up in writing. The contractor should ensure only the Network Manager, and not some designated representative, is the person notified. Either or both the contractor and the Network Manager should then contact members of the law enforcement community to determine additional efforts to identify and apprehend the perpetrator. Lost material should be identified by make, model number, serial number, color, and other identifying elements. The contractor may wish to notify their insurance carrier to make good on the loss. Copies of these reports should be provided to the Network Manager after the insurance carrier and the law enforcement community allows such documents to be made public.

Subcontractor Activities

Subcontractor activities normally end at the completion of milestones or phases. This is the usual reporting mechanism to inform the Network Manager of subcontractor activities. Many of the same types of reports supplied by the contractor may also be supplied by the subcontractor, depending upon the complexity of the project. In a project extending across multiple states, or even across several large cities within a state, subcontractors provide reports directly to the Network Manager, with information copies to the contractor.

Test

Test reports will include text, graphics, charts, tables, and even photographs. They will be voluminous, technical, and exhaustive in their detail. The schedule of reports, as well as the form and content will be mandated by the RFP and the resulting contract. The Network Manager should not allow the contractor to deviate from the required format and content of these reports. The contractor, or a third party who is independent, is usually charged with conducting the testing and recording the results.

Turnover

Reports on turnover of key personnel within the contractor's staff are of very high importance to the Network Manager. Many of these key personnel are also very senior and provide leadership as well as management of junior engineers and technicians. In extremely large procurements, the Network Manager may retain the right to participate in the selection of replacement when these senior people leave the contractor's organization. The contractor should provide the Network Manager with a report listing the name and date of departure of those senior personnel who are assigned to the Network Manager's project. This report should be submitted as soon as possible. As personnel matters are somewhat proprietary, the Network Manager must ensure such information is not made public.

These may not be the only reports which may be needed for a major project. The Network Manager must determine, in advance, what additional reports are needful and require their submission in the RFP or the contract.

Summary

1. This chapter introduces a few of the more salient points on project management for telecommunications networks.

2. We introduced the project management triangle with the relationships between time, money, and labor.

3. The Critical Path Method was touched upon very briefly as a method for creating Gantt charts and milestones. Those who are seriously interested in project estimation and formal project management need much more information than was presented here.

4. The classes of issues and the process for disposing of them was also presented. This issue work-off process is important and must be fully considered before the project begins.

5. And finally, the types and frequencies of reports were mentioned. It is important to note that these are the absolute minimum for any project and other, more customized forms and frequencies of reporting may be required.

Organization Description

The firm under study is the Bent Metal Corporation (BMC), Incorporated. They are purveyors of electronic chassis and sheet metal to the electronics and computer industry. The firm makes card racks, front panels, whole enclosures, and consoles.

BMC is physically located in two adjacent buildings in the local industrial park. It has been in existence for roughly seven years and is preparing to move from a terminal-mainframe approach to a distributed computing environment. The Network Manager works for "Big Mike" Iron, the director of MIS. Mike has read many recent articles in the trade press, has attended Network and Interop for the past few years, and feels the need to get one or more LANs with personal computers and the software that goes with them. Figure A-1 shows the present state of the firm. As we are not interested in too many details, the names of the workers and lower-level supervisors have been eliminated, leaving only their position titles.

What follows are some of the forms useful in capturing data about the existing and future information interchange needs of BMC or other like organizations.

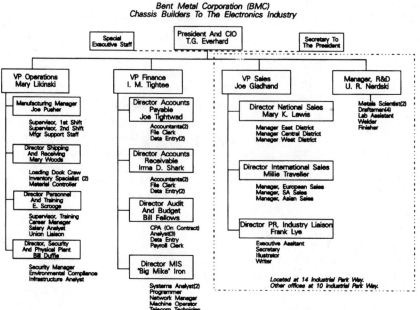

Figure A-1. Organizational chart.

The following forms are the ones to be used to obtain equipment inventory for the purposes of preparation for the RFP. This is bean counting of the most basic type, yet cannot be ignored as it is the basis for the entire design. This form is generic and may apply to many organizations.

Current Equipment Distribution Form

Name_____

Division_____Department_____Office_____

I have/share the following data communications equipment available:

(If none, check none, otherwise check all that apply)

None_____

Standalone Personal Computer_____Terminal only_____

Personal Computer on a LAN_____

Personal Computer emulating a terminal_____

Other (specify)_____

If you checked "Personal Computer on the LAN," please indicate what file server(s) you are connected to_____

I have a MODEM connected to my computer or terminal_____(Yes or No)

I have the following facsimile equipment available:

(Check all that apply)

A shared fax machine_____

Facsimile within my personal computer_____

Where is the shared fax machine located?

How often subscribers exchange data with other employees must also be captured to determine current needs. The minimum network throughput must meet these requirements.

Current Telecommunications Frequency Form

Please list the Division, Department and Office you have data or facsimile telecommunications with and how often you contact them. If none, check the line "none."
None_____

Division   Department   Office   Frequency*

*H = Hourly, D = Daily, W = Weekly, M = Monthly.
NOTE: There may be several different frequencies for telecommunications within the same group. You may send a daily as well as a monthly summary to the Director Of National Sales. Please enter both.

Current Internal Paper Communications Frequency Form

Please list the Division, Department and Office you send paper-based documents and messages to and how often you do this.

Division   Department   Office   Frequency*

*H = Hourly, D = Daily, W = Weekly, M = Monthly.
NOTE: There may be several different frequencies for paper-based communications within the same group. You may send a daily as well as a weekly summary to the Director Of Audit & Budget. Please enter both.

Much of the existing paper based information interchange can be replaced by electronic formats. The purpose of the form shown below is to capture the paper-based information flow.

Current External Paper Communications Frequency Form

Please list the organization you send paper-based information to and how often you do this.

Organization Name And Address   Frequency*

*H = Hourly, D = Daily, W = Weekly, M = Monthly.

Many times current subscribers, and those not connected to the networks, are quite sophisticated in their understanding of commercial, off the shelf software. They will be a good source of information to the Network Manager. The form below is one method used to capture software needs.

Future Software Needs

Assume you have a personal computer on your desk, and the personal computer is connected to a local area network (LAN). Please check the type of software you will need to properly complete your assigned duties.

_____Word Processing	_____Spreadsheet	_____Graphics
_____MODEM Communications	_____Publishing	_____Accounting
_____Statistical Processes	_____Project Mgmt	_____CAD/CAM
_____Data Base	_____Payroll	_____Sales Tools
_____Case Tools	_____Terminal Software	_____Tetris
_____Other(s) List:		

Figure A-2 is still another method of capturing information interchange between subscribers within an organization. This is not the only approach, but it is one we have used previously, with some success.

Directions

The matrix in Fig. A-2 is completed along with the forms above concerning the exchange of information between yourself and others in the firm. Down the left-hand column list the division, department, and office. Across the top note the time of transmission of the paper or electronic message. Note that each intersecting row and column is split in half diagonally. The top left half is for number of pages of paper, the bottom right half is for the size of the file transfer.

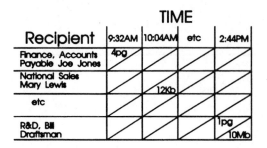

Figure A-2. Information interchange matrix.

Explanation of This Form

Although this matrix is a small sample of the actual interconnectivity, the Network Manager will find the final product to show the user's estimate of the amount of information exchanged on a minute-by-minute basis. The completed matrix will provide, by horizontal and vertical summing, the point-to-point traffic within the organization. This will give the Network Manager the backbone throughput required.

The need for paper-based information interchange may not be obvious. Once the LAN is fully functional, much of the paper-based information interchange will be replaced by electronic methods. A rule of thumb is that one page of single-spaced text is equal to 12,000 bits of data. Therefore, by multiplying the number of pages by 12,000, the

Network Manager can find the total estimated throughput, in bits per second, required.

Other telecommunications such as modem-based communication outside of the organization or terminal to mainframe telecommunications within the organization are somewhat harder to estimate. We recommend a final "fudge factor" in throughput of roughly 20 percent when the LAN design is completed. If the mathematics indicate a throughput of eight megabits per second is required, the Network Manager may consider something other than a single IEEE 802.3-based LAN. Multiple IEEE 802.3-based LANs separated by routers may be useful.

A very key point must be brought forward. The Network Manager is not a network designer by trade or profession in most circumstances. The end result of this effort is a set of specifications on which vendors can bid and build toward. The Network Manager must have enough data on hand to look at a vendor's response and decide whether the response is anywhere near what is needed. This is not an exercise in sizing the end product; rather, it is an attempt to build a set of boundaries for others to fit within. Much of this exercise is "bean counting" and seems quite useless. It is not; those persons designing the new network may or may not need this information. It is better that they have it, than if they do not. For example: How many facsimile machines are in use? Can these be replaced by facsimile cards in PCs and laser printers? By identifying the location of the existing facsimile machine, the designer knows which new PCs must have facsimile cards in them.

Bent Metal Corporation Chassis Builders to the Electronic Industry

An RFP For Network Specification, Design, and Installation

New Way Industrial Park
Progress, New Hampshire 03999

Definitions of Words Used

ATM: An acronym for Asynchronous Transfer Mode, a high-speed data communications protocol.

Bidder: A person or firm who responds to this request for quotation and is subsequently awarded the job through a contract.

BMCNet: Bent Metal Corporation Network. The products to be installed and tested as described in this request for quotation.

Contract: A binding, legal document that details the agreement between Bent Metal Corporation and the successful Bidder.

Data communications: Any binary telecommunications information coming from or going to a computer.

EIA/TIA: Electronics Industry Association/Telecommunications Industry Association

FDDI: An acronym for Fiber Distributed Data Interface, a high-speed data communications protocol.

Frame relay: A medium- to low-speed telecommunications protocol.

IEEE 802.3: The nonproprietary specification provided by the Institute Of Electrical and Electronic Engineers for a high-speed, packet-based LAN protocol.

NM: Nano meters. 1^{-9} meters, a unit of measurement for light waves.

OTDR: An acronym for Optical Time Domain Reflectometer, a device to provide visual indications of the physical condition of fiber-optic cable.

Response: A Bidder's documentation answering this request for quotation.

ST: One of several types of termination used with fiber-optic cable; the only one acceptable in this request for quotation.

Terms and conditions: The technical and legal specifications in this request for quotation.

UTP: Unshielded twisted pair, a specific type of wire used for transmission media.

Video: Any moving graphic image.

Voice: Human speech prior to digitization.

Request for Quotation*

I. Bid Instructions

A. Aim of this RFQ

1. This RFQ is to provide Bidders with information to enable them to prepare a cost effective and technically sufficient response.
2. This RFQ establishes some specific and general parameters for a network which will meet the high speed data, voice, and video telecommunications needs of BMC.
3. This RFQ is issued for the facility located at: Both 10 and 14 Industrial Park Way, Progress, NH 03999

B. RFQ Schedule. The schedule for this RFQ is as follows:
RFQ Issued 9 March, 1995
Bidder visits 21 and 22 March, 1995. Bidders are allowed two hours only.
Quotations submitted 15 April, 1995
C. Bidder inquiries.

1. If additions, deletions, modifications, or clarifications to this RFQ become necessary, Bidders will be notified in writing at the address given in their responses.
2. Inquiries concerning this RFQ should be directed to:
Frank Jones, Network Manager
Bent Metal Corporation
MIS Department
10 Industrial Park Way
Progress, NH 03999
(603) 123-4567

D. Response submission.

1. Three (3) copies of the quotation must be submitted to the person and address shown in C2 above.
2. The responses should be submitted on or before: 15 April, 1995.

---

*This document is NOT legally sufficient as it stands. The reader must seriously consider obtaining legal guidance before putting their RFP in the public domain.

a. Responses not received by close of business this date will be automatically disqualified from consideration.

b. All material received in response to this RFQ shall become the property of BMC and will not be returned to the Bidder. Regardless of the Bidder selected, BMC shall reserve the right to use any information presented in a response.

3. Each Bidder must prepare a quotation which follows the format of the one shown in this RFQ. Any exceptions to this RFQ should be explained in as much detail as possible. All pages of the quotation must be consecutively numbered.

4. Each item presented in this RFQ must be responded to thoroughly. Failure to address any of the requirements will be grounds for a Bidder's automatic elimination from the Bidder list.

5. If the response does not meet the requirements shown, the Bidder must explain why such a discrepancy exists.

6. The response will be signed by someone in the firm who is authorized to bind the firm and negotiate with representatives of the Bent Metal Corporation (BMC).

E. Basis of award.

1. BMC will award the Contract based on the following characteristics.

a. Technical elegance combined with cost efficiency.

b. Most closely matching the specifications set forth in this RFQ.

c. The quality of references provided by the Bidder.

d. Timeliness and responsiveness of responses from the Bidder.

e. Technical qualifications of the Bidder.

2. BMC declares that it reserves the right to accept or reject any or all responses, to take exception to these RFQ specifications, or to waive any formalities.

3. Bidders may be excluded from further consideration for failure to comply with the specifications of this RFQ.

4. A Bidder's quotation does not place BMC under any obligation to accept any response based on the lowest priced or most technologically advanced design.

5. Some of the criteria will be used in evaluating the responses to this RFQ include:

   a. Reliability of the network and Bidder—Bidder must have an acceptable rating to be considered by the BMC.
   b. Ability to meet all BMC requirements.
   c. Bidder's capabilities to provide 24 hour a day response to problems.
   d. Warranty/guarantee length and wording will be considered above and beyond those items listed in E.1.e above.
   e. BMC will give preference to Bidders who actively work for future enhancements in telecommunications networks and have some presence on standards committees.

6. Selection, notification and award of contract

   a. BMC shall reserve the right to clarify the terms and conditions of any response submitted.
   b. If the BMC awards a Contract relative to this RFQ, the successful Bidder shall be advised by letter. A Contract shall be developed by BMC and shall incorporate in its provisions this RFQ and the successful response.
   c. Public announcements or news releases pertaining to this RFQ shall not be made without the written permission of BMC.
   d. All respondents will be notified whether their response has been accepted or not, no later than: 1 May, 1995

F. Duration of quotation. The price noted will be in effect for a period of ninety (90) days after the signing date of the Contract which springs from this RFQ or 15 April, 1995, whichever is later.
G. Contractual relationship. The contents of the quotation and this RFQ are to be considered part of the Contract between the BMC and selected Bidder.
H. Slippage of schedule.

1. When the selected Bidder knows of problems that can possibly delay the completion of the effort, the Bidder will notify the person(s) noted below providing them with reasons why the slippage will occur, the new completion date and other choices which may be taken.
   Frank Jones, Network Manager
   Bent Metal Corporation

MIS Department
10 Industrial Park Way
Progress, NH 03999
(603) 123-4567

2. Even if the Bidder completes the notification as noted in H.1 above, they are not relieved of responsibility. If the Bidder fails to notify as in H.1 above, this will become a basis for determining whether the Bidder is negligent or not.

I. Laws, ordinances and codes. All applicable federal and state laws and the rules, regulations, codes, and ordinances of all local authorities having jurisdiction will apply to the Contract throughout. They will be deemed to be incorporated in the Contract.

J. Technical elegance.

1. Any Bidder who finds a more technically elegant way to complete the required design may take exception to this RFQ. These exceptions will be fully documented showing what changes are recommended, and what cost savings are created. Substitutions of one material for another is not a technically elegant change. BMC will provide written authority to perform such changes.

2. All materials provided by the Bidder will meet the requirements shown in the appendixes to the RFQ.

K. Other terms and conditions.

1. Bidder should make every attempt to use terminology in their response that is defined in the RFQ. If other words, phrases or descriptions are in use, definitions should be included. In case of conflicts, those definitions noted in this RFQ will apply.

2. Bidder shall include a recent financial statement and proof of insurance coverage with their quotation.

3. BMC will not be liable for any costs incurred in the preparation and presentation of the quotation.

4. Any material submitted by the Bidder that is considered confidential in nature must be clearly marked as such and submitted under separate cover.

L. Reliance. The Bidder acknowledges that they are, and that BMC relies upon, the Bidder as an expert, fully competent in all phases involved in producing, testing, developing, installing, modifying, altering, servicing, and integrating the equipment and or systems furnished in the telecommunications network. In this context, the Bidder agrees that it will not deny any responsibility or obligation to BMC on any grounds. More specifically, and without limiting the above, BMC in originating, furnishing, or approving any specification, drawing, plan, change, schedule or other document or part thereof, or any test report, or in accepting any networks, neither accepts responsibility for nor relieves the Bidder from the performance of all terms and conditions of this RFQ and any Contract that may be awarded. Any such acts by BMC shall not modify, impair, or abrogate any rights of BMC under this RFQ and any subsequent Contract.

II. Objectives of this RFQ. To provide a clear description of the networks, applications, connectivity, and hardware which will be used to support high speed data, voice, and video telecommunications. This new installation will be called BMCNet in the remainder of this document.

III. Distribution network requirements. A successfully completed installation is the goal of the BMC. Therefore the Bidders must meet the following requirements.

A. Overall Requirements. BMCNet be capable of supporting 10 Mbps IEEE 802.3, 100 Mbps FDDI, 150 Mbps Asynchronous Transfer Mode (ATM), Frame Relay, telephone service and internal videoteleconferencing.

   1. BMCNet must have a minimum four (4) year useful life without major modifications.
   2. All products/equipment/workmanship MUST be fully compliant with FDDI (both single AND dual attached systems) TR 48-1, and EIA/TIA Standard 568 and 569. In case of contradiction or conflict between standards, BMC will select the appropriate one(s) to be used.
   3. BMCNet is to be built on requirements described in this RFQ, its appendixes and the supplied blueprints and drawings.

B. Architecture/topology.

   1. BMCNet must be based on an open distribution architecture so that existing equipment and facilities as well as future equipment from other Bidders can be supported.

2.   Topology is a hierarchy of stars.

C. Transmission medium. Bidder will install wire and cable that meet or exceed the specification described in the EIA/TIA specifications noted in III.A.2 above.

D. Transmission media design.

1. Where inner duct is run to carry fiber-optic cable, a separate pull string will be run through the duct so additional fiber can be pulled at a later date.

2. All new riser cable will be placed according to laws, ordinances, and codes. Bidder may use existing core drills, sleeves, or ductwork, or make provisions for new core drills. Authorization for such usage will be granted, in writing, by the BMC. Conduit inner duct MUST be installed for all fiber-optic cable. This requirement may be waived where the fiber-optic cable is brought out to the equipment to which it connects.

3. The physical design of all transmission media will not include splices.

4. At each place where fiber optic cable connects to a piece of telecommunications equipment or a fiber optic patch panel, three (3) meters of slack will be left to facilitate patching and relocation.

5. Three (3) to four (4) meters of slack will be left in all fiber-optic runs to facilitate splicing at a later date by persons who may not be a member of the Bidder's staff.

6. The Bidder will abide by the cable manufacturer's specifications for bend radius, tensile rating, and maximum vertical rise.

7. When pulling runs of cable longer than 200 meters through duct or conduit, less than 40 percent fill ratio is mandated.

8. Bidder will fire stop all holes in the riser/telecommunications closet or any additional fire wall openings which accommodate the placing of the cable. Acceptable fire stop products must be equivalent to 3M fire stop products and meet all local code requirements.

9. Materials to be supplied by Bidder may include, but are not limited to that shown in the Bidder provided bill of materials.

10. The Bidder work functions include the following work items for the installation of the system:

   a. Placing riser cables as required.

b. Terminating all fiber strands using ST connectors with ceramic inserts.

c. All fiber and copper runs are installed unspliced.

d. Placing conduit inner duct as required.

e. Connecting coaxial cables to self-terminating connectors.

f. Terminating UTP cable with plugs or receptacles as required.

g. Providing labels and marking termination hardware.

h. Install all floor and wall-mount racks where required.

i. Mount equipment in racks, then connect cable to equipment.

j. Leaving new pull strings in all conduits, ducts, or sleeves where cable has been pulled under this Contract. The Bidder will ensure this specification is not in contradiction to laws, ordinances, and codes.

11. All copper cable runs that terminate within a wall, under a raised floor, or in under-floor outlet boxes will have 0.5 meters (19 inches) of slack left in the cable.

E. Fiber-optic termination method. Full termination of all fiber strands is required at the time of installation. Fiber that is unused is to be capped and left dark. Bidder will supply patch cables for each pair of ST connections. Patch cables will be checked for loss and reflection.

F. Telecommunications rooms and closets.

1. All exposed fiber-optic strands will be protected from dirt, dust, and moisture.

2. Bidder will base this response on the following factors in their given order of precedence: Human safety, reliability, expansibility, flexibility, telecommunications security, space utilization, and human factors.

3. Layout, including rack locations, will be submitted for approval prior to the start of installation.

G. Wire and cable numbering schemes. Numbering schemes will be derivative in nature, the first digits deriving from the device the wire or cable is attached to. The final numbering scheme will be submitted for approval prior to the start of installation.

H. Start/completion dates. To be negotiated.

IV. Installation, test, and acceptance.

A. BMC's Responsibilities. BMC will:

1. Allow the Bidder's employees free access to the premises and facilities at all reasonable hours during the installation.
2. Take actions as necessary to ensure that premises are dry and free from dust and in such condition so as not to be hazardous to the installation personnel or the material to be installed.
3. Provide heating and general illumination in rooms in which work is to be performed or materials stored. BMC expressly waives any responsibility for any material, tools, or equipment left on the premises after the end of the normal work day.
4. Furnish adequate detailed drawings of the building to allow installation of cables and equipment by the Bidder.
5. Provide free and clear access to existing conduit or the placement of new conduit if necessary to all work locations, floors, buildings, and etcetera to support the cable installation and provide Bidder access to these adjacent areas where and when required.
6. Provide 110 volts, 20 amp, 60 Hz commercial power necessary for the installation and for future equipment needs.
7. Make alterations and repairs to the building, equipment, or services if it is determined by BMC to be desirable or necessary for safe operation.
8. Make inspections when notified by the Bidder that the equipment or any part thereof is ready for acceptance.

B. Bidder responsibilities. The Bidder will:

1. Provide all supervision, labor construction, tools, equipment, materials, transportation, erection, storage, construction, unloading, inspection, inventory keeping, and returning spare material as specified in the attached contractual documents. Whenever in this RFQ the terms provide, furnish, supply, install, etcetera are used, it shall be interpreted as requiring the Bidder to both furnish and install materials, unless specific provisioning and/or installing of the materials by BMC is denoted.
2. Obtain BMC's permission before proceeding with any work necessitating cutting into or through any part of the

building structure such as girders, beams, concrete, tile floors, or partition ceilings.

3. Be responsible for and promptly repair all damage to the building due to carelessness of its workmen and exercise reasonable care to avoid any damage to the building. Report to BMC any damage to the building which may exist or may occur during the occupancy of the quarters.

4. Take necessary steps to ensure that required fire fighting apparatus is accessible at all times. Flammable materials shall be kept in suitable places outside the building(s).

5. Install all hardware in accordance with manufacturer's specifications or local codes and ordinances, whichever is more stringent.

6. Conduct certain tests and record the results. These are more fully explained in paragraphs C through G below.

7. Promptly notify BMC of completion of the work on equipment of such portions thereof that is ready for inspection.

8. Give BMC notice of intended completion of the installation at least one week prior to completion.

9. Promptly correct all defects for which the Bidder is responsible as determined by BMC.

10. Coordinate all work with those persons noted in section I.C.2 above, or their designated representative(s).

11. Maintain insurance on BMCNet until such time as it is accepted by BMC.

12. Remove all tools, equipment, rubbish, and debris from the premises and leave the premises broom clean and neat upon completion of the work.

13. The Bidder may use subcontractors to perform work. However, all responsibilities rest with the Bidder, and the Bidder must furnish the complete list of such subcontractors. Subcontractors are required to carry the same insurance protection as that in force by the Bidder.

14. The Bidder must abide by the safety and security rules in force on the work site.

C. Test/acceptance criteria.

1. The response must contain a statement accepting the test criteria as shown in this RFQ.

2. Once the work has been completed the Bidder will provide drawings (redlines) showing where cables are run,

the location of patch panels, splices, or other like installation hardware.

D. Performance measurements and requirements.

1. All fibers with their terminating connectors are to be measured and recorded in both directions for attenuation. Optical Time Domain Reflector (OTDR) traces are to be recorded to verify the uniformity and integrity of the installed fiber. Attenuation measurements are to be acquired using power meter and optical power source at the appropriate wavelengths for the type of fiber under test. Return loss measurements may be accomplished with return loss measuring sets or combined OTDR readings if the OTDR is so equipped.

2. All fiber-optic measurements are to be accomplished using test jumpers, no less than 100 meters long at both ends of the fiber under test. All recorded measurements are the average of three stable readings.

3. Measurements and performance requirements are noted in EIA/TIA Standard, FOTP-107, Return Loss for Fiber Optic Components, EIA/TIA-455-107.

E. Measurement equipment requirements

1. Capable of measuring attenuation and return loss at 820, 850, 1300, and 1550 nm.

2. Capable of generating center wavelength of 820 ± 10 nm, 850 ± 10 nm, 1300 ± 20 nm and 1550 ± 50 nm.

3. Capable of maintaining source spectral bandwidth of 820, 850, 1300, and 1550 nm at ± 2.2 percent accuracy.

4. Capable of maintaining accuracy of ± 0.05 dB for measurement of attenuation.

5. Capable of maintaining accuracy of ± 1.0 dB down to −50 dB for measurement of return loss.

F. Fiber performance measurement

1. Bidder or supplier will certify that the cable provided meets the specifications noted.

2. OTDR testing is performed and recorded AFTER installation and connectorization.

3. Tests will be performed on one hundred (100) percent of all fiber runs installed. Test results will be documented.

G. Unshielded twisted pair (UTP) level-five test procedures.

1. The following tests are conducted on a Go/No Go basis. Failure of any test is a failure of that run. The run will be repaired or replaced. The use of splices as a repair activity is strictly prohibited.

   a. Shorts within a pair.
   b. Opens within a pair.
   c. Pair reversals.
   d. Pair transpositions.

2. Tests will be conducted on 100 percent of all UTP Level Five runs installed.

V. Training and documentation.

A. Bidder's technician training must include such items as ST, FDDI, and ATM termination practices. Cable loss budget analysis, testing procedures, test equipment operation, and documentation must also be part of the curriculum.

B. Complete diagrams of the final cable runs, showing patch panels, color coding, numbering schemes, and cross-referencing as required.

C. Bidder will be prepared to provide calibration certificates for their equipment used for these tests. Calibration will be performed in accordance with the test equipment manufacturer's specifications.

VI. Financial.

A. The Bidder's response must include complete pricing for the described data telecommunications network. Pricing will be for a turnkey operation. Pricing will not reflect material provided by BMC.

B. The total will include cost for materials, labor, transportation, sales/use taxes, maintenance, and miscellaneous charges must be separately identified.

C. Complete materials lists and unit prices must be included.

D. A statement of estimated total labor hours and prevailing labor rates must be included.

E. Should certain additional work be required, or should the quantities of certain classes of work be increased or decreased from those upon which the Bidder's response is based, only by order of approval by BMC, the undersigned agrees that supplemental unit prices shall be the basis of payment to the Bidder or credit to BMC for such additional work, or increase or decrease in the work. Unit prices shall represent the exact net amount per unit to be paid to the Bidder (in case of additions) or credited to BMC in case of decreases. Overhead and profit are not included in unit prices and no additional adjustments will be allowed for overhead, profit, insurance, compensation insurance, or other direct or indirect expenses of the Bidder, subcontractor, or other supplying parties.

VII. Additional requirements.

A. Previous projects.

1. Have been completed within the last two (2) years.
2. Include customer name and contact telephone number.
3. Preferably be within the same metropolitan area as BMC facility or within a 100-kilometer radius.

B. The following background information on your company should be provided:

1. How long has your company been in the cable installation business?
2. How many meters of fiber-optic cable have you installed?
3. How many ST type connectors have you installed?
4. How many FDDI type connectors have you installed?
5. How are your technicians/installers trained or certified?

VIII. Bidder acceptance form.

This form or a reproduced copy will accompany the response. All forms must be signed in original.

BIDDER ACCEPTANCE FORM
Name Of Authorizing
Officer_____
Title_____
Name Of
Firm_____
Address_____
Telephone_____
City, State & Zip Code_____
I, _____ accept the technical specifications as outlined in this RFQ for BMC and am proposing an appropriate high speed data voice and video telecommunications network which will meet BMC's mandatory requirements or have clearly noted any exceptions. It is my understanding that my Response, if accepted, will become a part of the Contract and that the quoted prices will remain in effect for ninety (90) days from the bid closing date.

_____   _____
Bidder Authorized Signature   Date

Appendix **C**

Contract for Network Installation

Contract for Network Installation*

This Agreement (the "Agreement") is made this 15th day of April, 1995, by_____("Contractor") and Bent Metal Corporation, "BMC" a New Hampshire corporation.

1. TYPE OF CONTRACT

     This Agreement is for the installation of a teleecommunications distribution system which will meet the high speed data, voice, and video telecommunications needs of the Bent Metal Corporation

2. DESCRIPTION OF THE WORK

     A. Contractor shall perform certain work (the "Work") as shall be specified by BMC in writing during the period beginning on_____and ending on or before_____(the "Term"). The Work shall be described in an RFQ which is included as part of this contract by reference.

     B. A purchase order shall be executed for the Work. Contractor shall be bound by the provisions of the purchase order as to the Contract Sum, the Work Site description, and other conditions of the Work, regardless of whether Contractor acknowledges or otherwise signs the purchase order, unless Contractor objects to such terms in writing.

     C. The Work shall be performed in accordance with the RFQ attached hereto as Exhibit A and incorporated herein by reference. All terms which are defined in this Agreement and used in Exhibit A shall have the same meaning in the Exhibit as in this Agreement.

3. TIME OF COMPLETION

     A. This Agreement shall be in effect from_____ through _____(the "Term").

     B. A Project shall not be deemed complete until accepted in writing by BMC.

4. CONTRACT SUM

     A. BMC shall pay Contractor, for the performance of this contract such amounts as may be specified in the purchase order.

*This document is NOT legally sufficient as it stands. The reader must seriously consider obtaining legal guidance before using it as a tool to bind respondents to the RFP.

B. The Contract Sum includes:

(1) The cost of all labor, equipment, materials, tools, supplies, permits, and licenses necessary for or incidental to the completion of the Work that are identified only in the response to the RFQ; and

(2) All contributions, assessments, taxes, fees, and other costs related to the Work that are specified in the response to the RFQ.

C. Contractor acknowledges that the Contract Sum is based on Contractor's inspection of the site specified in the RFQ where the Work is to be performed (the "Work Site").

D. No allowances shall be made to Contractor for extra costs as a result of difficulties encountered during the Work.

E. All materials incorporated in the Work shall become the property of BMC upon payment for the Work.

5. WORK SITE

A. Contractor shall perform all Work without interference to BMC's employees or operations in areas around the Work Site.

b. Contractor shall keep the Work Site clean and free from rubbish on a daily basis. Upon completion of the Work Contractor shall remove all waste materials, tools, and materials from the Work Site and leave the Work Site "broom clean."

6. SUPERVISION

A. Contractor shall be solely responsible for:

(1) All means, methods, techniques, sequences, and procedures used in the Work; and

(2) The acts and omissions of all agents and employees of Contractor, all subcontractors of Contractor and their agents and employees, and all other persons performing any part of the Work.

B. Contractor shall enforce strict discipline and good order among its employees at all times. Contractor shall not employ any person unfit or unskilled in that portion of any Work assigned to him.

7. WARRANTY

A. Contractor warrants to BMC that:

(1) All materials and equipment incorporated in the Work shall be new; and

(2) The Work shall be of good quality and workmanship, free from faults and defects, and in conformity with the requirements of this Agreement, including any plans or specifications incorporated in this Agreement.

B. The warranty specified shall be in effect for a period of one year from the date of BMC's written acceptance of the Work (or such longer period as may be prescribed by law).

C. If any part of the Work is defective or otherwise not in conformity with the requirements of this Agreement, BMC shall have the right to require prompt correction at the sole expense of the Contractor. Payment by BMC of the Contract Sum shall not relieve Contractor of the obligations specified in this article.

8. INVOICING AND PAYMENTS

A. Contractor shall invoice BMC as follows:

B. Each invoice shall set forth the amount due from BMC to Contractor, in accordance with the provisions of article 4 above, for Work performed.

C. BMC reserves the right to withhold, reduce, or recover payment of any portion of the Contract Sum if Contractor fails to pay when due any third parties for labor, materials, or other costs incurred by Contractor in the performance of any Work.

D. If Contractor is not in default under this Agreement, BMC shall pay each invoice within 45 days after BMC's receipt of the invoice, subject to the provisions above.

E. Contractor's acceptance of that portion of the Contract Sum invoiced shall constitute a waiver of all claims by Contractor with respect to the Work covered by the invoice.

9. CHANGES

A. BMC may order changes in any Project at any time. In the event of any such change, the Contract Sum shall be adjusted accordingly. Any such change shall be documented on

BMC letterhead and signed by an individual who is authorized to bind BMC to contractual requirements.

B. Except as provided above, any claim by Contractor for additional compensation arising out of this Agreement must be made in writing within ten (10) days after the commencement of the event giving rise to the claim. Otherwise, any such claim shall be deemed waived by Contractor, notwithstanding actual notice thereof on the part of BMC.

10. RELATIONSHIP OF PARTIES

A. The relationship of BMC and Contractor is that of owner and independent contractor, and not that of master and servant, principal and agent, employer and employee, partners, or joint venturers.

B. BMC may let other contracts in connection with the Work. Contractor shall actively cooperate with any other contractors retained by BMC.

C. Contractor shall neither delegate any duties nor assign any rights under this agreement without the prior written consent of BMC. Any such attempted delegation or assignment shall be void. This does not prohibit the Contractor from subcontracting part(s) of the Work. BMC shall be informed of the subcontracting pertinent to the Work.

D. Neither Contractor nor Contractor's agents, employees, or subcontractors shall disclose to any person or entity any confidential information of BMC, whether written or oral, which Contractor or Contractor's agents, employees, or subcontractors may obtain from BMC or otherwise discover in the performance of any Work. As used in this section, the term "confidential information" shall include, without limitation, all information or data concerning or related to BMC's products (including the discovery, invention, research, improvement, development, manufacture, or sale of BMC products) or general business operations (including sales costs, profits, pricing methods, organization, employee lists, and processes).

11. SAFETY AND SECURITY

A. Contractor shall provide adequate protection for its equipment and all Work, and shall provide such suitable safety appliances as may be needed to safely perform all Work.

B. Contractor shall perform all Work under, and shall ensure that all of Contractor's employees and agents engaged in any Work also operate under stringent safety precautions. Contractor is familiar with, and shall abide by, BMC's safety procedures in effect at the Work Site as well as safety procedures generally applicable to persons on BMC's premises.

C. Contractor shall perform all Work in accordance with BMC's security procedures in effect at the Work Site.

D. The provisions of this article shall:

(1) Apply to all procedures of BMC, whether posted or otherwise, and shall include policies and guidelines as well;

(2) Not limit or otherwise affect in any manner Contractor's obligations pursuant to all articles of this Agreement; and

(3) Not impose any obligation upon BMC to enact or enforce specific safety or security procedures, policies, or guidelines.

12. APPLICABLE LAWS

A. Contractor shall comply with all applicable federal, state, and local laws, rules, regulations, or orders issued by any public authority having jurisdiction over the Work, including without limitation:

(1) The Williams Steiger Occupational Safety Health Act of 1970, as amended, and any rules, regulations, or orders issued thereunder; and

(2) All applicable nondiscrimination requirements, including without limitation the provisions of Presidential Executive Order 11246 and the rules and regulations issued thereunder.

B. This Agreement shall be construed in accordance with the laws of the State of New Hampshire.

13. INSURANCE

A. During the Term and at all times that Contractor performs services for BMC, Contractor shall maintain in full force and effect, at Contractor's own expense, insurance coverage to include:

(1) WORKERS' COMPENSATION AND EMPLOYER'S LIA-
BILITY INSURANCE

Workers' Compensation insurance shall be provided
as required by law or regulation.

Employer's Liability insurance shall be provided in
amounts not less than $500,000 per accident for bodily
injury by accident, $500,000 policy limit by disease, and
$500,000 per employee for bodily injury by disease.

Where permitted by law, such policies shall contain
waivers of the insurer's subrogation rights against BMC.

(2) GENERAL LIABILITY INSURANCE

Contractor shall carry either Comprehensive General
Liability Insurance or Commercial General Liability
Insurance with limits of liability and coverage as indi-
cated below:

(a) Premises and Operations.
(b) Products and Completed Operations.
(c) Contractual Liability.
(d) Broad Form Property Damage (including Completed
Operations).
(e) Explosion, Collapse, and Underground Hazards
when Contractor will create risk normally covered
by such insurance.
(f) Personal Injury Liability. Comprehensive General
Liability policy limits shall be not less than a
Combined Single Limit for Bodily Injury, Property
Damage, and Personal Injury Liability of $1,000,000
per occurrence and $1,000,000 aggregate.

Commercial General Liability (Occurrence) policy
limits shall be not less than $1,000,000 per occurrence
(combined single limit for bodily injury and property
damage), $1,000,000 for Personal Injury Liability,
$1,000,000 Aggregate for Products and Completed
Operations, and $2,000,000 General Aggregate.

Except with respect to Products and Completed
Operations coverage, the aggregate limits shall
apply separately to Contractor's Work under this
Agreement.

Such policies shall name BMC, its officers, direc-
tors, and employees as additional Insureds and

shall stipulate that the insurance afforded Additional Insureds shall apply as primary insurance and that no other insurance carried by any of them shall be called upon to contribute to a loss covered thereunder.

If "claims made" policies are provided, Contractor shall maintain such policies, without endangering aggregate limits at the above stated minimums, for at least five (5) years after the expiration.

(3) AUTOMOBILE LIABILITY INSURANCE

Contractor shall carry bodily injury, property damage, and automobile contractual liability coverage for owned, hired, and non-owned autos with a combined single limit of liability for each accident of not less than $1,000,000.

(4) CERTIFICATE OF INSURANCE

Certificates of Insurance evidencing the required coverages and limits shall be furnished to BMC before any work is commenced hereunder and shall provide that there will be no cancellation or reduction of coverage without thirty (30) days prior written notice to BMC. All insurance policies shall be written by a company authorized to do business in the State of New Hampshire. Contractor shall furnish copies of any endorsements subsequently issued which amend coverage or limits.

14. INDEMNIFICATION

A. Contractor shall defend, indemnify, and hold harmless BMC from any and all claims, losses, demands, attorneys' fees, damages, liabilities, costs, expenses, obligations, causes of action, or suits:

(1) For injury (including death) to any person or damage to or loss of any property arising out of or resulting from any act or omission, whether active or passive and whether actual or alleged, of Contractor or its employees, agents, or subcontractors, to the maximum extent permitted by law;

(2) Arising out of or resulting from Contractor's breach of this Agreement; or

(3) Arising out of or resulting from labor, materials, services, or supplies furnished by subcontractors or suppliers of Contractor and from all related liens (including without limitation laborers', materialmens', or mechanics' liens).

B. Contractor shall promptly notify BMC in writing of any matter as to which Contractor is bound by the indemnity in section 14.A above, and shall do all things required to protect BMC's interests pursuant to this article.

15. TERMINATION

A. If Contractor neglects to prosecute the Work properly or fails to perform any provision of this Agreement, BMC may, after seven days' written notice to Contractor and without prejudice to any other remedy BMC may:

(1) Terminate this Agreement and take possession of all materials, tools, and equipment pertaining exclusively to the Work;

(2) Finish such Work by any means as BMC sees fit, or otherwise make good the deficiencies; and

(3) Deduct the costs incurred by BMC from any payment due Contractor.

B. If the unpaid balance of the Contract Sum exceeds the expense of finishing the Work or making good the deficiencies pursuant to sections above, BMC shall pay the difference to Contractor within ten days after completing the necessary Work. If such expense exceeds the unpaid balance, Contractor shall pay the difference to BMC upon demand.

16. NOTICES

Notices provided in connection with this Agreement shall be in writing and sent by certified or registered mail, return receipt requested, postage prepaid. Notices shall be addressed as follows:

Frank Jones, Network Manager
Bent Metal Corporation
MIS Department
10 Industrial Park Way
Progress, NH 03999

17. ENTIRE AGREEMENT

A. This Agreement represents the entire agreement of the parties with respect to the Work, and may not be amended except by a written instrument signed by both Contractor and BMC.

B. If any proposal or other documentation from Contractor is attached to this Agreement or otherwise referred to in this Agreement, this Agreement shall control to the extent that such proposal or documentation conflicts with this Agreement.

IN WITNESS WHEREOF, the parties have executed this Agreement as of the day and year first written above.

Bent Metal Corporation   Contractor

By:_____   By:_____

Index

Acceptance form, 100
Accurate floor plans, 96
Additional obligations, 97
Alternate routing, 30
Analysis of the data, 15
Application Layer, 62
Appropriate safety precautions, 146
Assumptions (by vendors), 111
Automatic switchover (vs manual), 32

Backup media, 31
BAFO (best and final offer), 113
Baseline:
  future performance, 81
  network management, 157
Basis for Award, 90
Bid bond, 121
Bridge:
  bridging routers, 73
  brouters, 73
Bridging of teams, 173
Bubble chart, 20
Budgetary skills, 3

CATV (as transmission media), 153
Changes (in company size) downward, 35
Changes (in company size) upward, 35
Client-server, 69
Code enforcement officers, 12
Codes (construction), 4
Codes enforcement, 97
Cold solder joint, 53
Common law, 123
Conflict resolution, 120
Construction foreman, 142
Construction loan, 120
Contract:
  and the RFQ, 84

Contract (*Cont.*):
  review by legal practitioners, 118
Contractor's employees, 141
Corporate culture, 171
Cost of manpower, 109
Cost overrun, 110
Critical path, 164
CSMA/CA, 70
CSMA/CD, 70

Danger to human life, 146
Data gathering, 15
Data Link Layer, 61
Definition of terms, 89
Depth of understanding by vendor, 113
Derivative numbering approach, 94
Derivative numbering systems, 96
Design (four step process), 9
Design and installation can change, 120
Distributed computing:
  as a label, xiv, 2
  not client server, 69
DIX standard, 50
Downsizing, 2
Duration of Quotation, 90

Easy mistake (for the manager), 46
Ego:
  a basis for problems, 132
  and lack of ability, 142
Enforceable provisions (contract), 117
Equipment substitutions, 114
Escrow, 120
Estimates of the duration, 165
Ethernet versus IEEE 802, 3, 63
Excess material, 88
Experts, 10
Exposure to the elements, 139

Face value (contract), 122
Feedback loops (for design), 6
File transfer test, 79
Final authority, 142
Final offer, 103
Financial stability (vendor), 99
Fire, theft, vandalism (insurance), 125
Floating grounds, 52
FOIRL (Fiber Optic Interface Relay), 152
Forced sale, 122
Four Step Process, 9
Fudge factor (in estimates), 142
Future protocols, 36

Gantt charts, 161
Gateway:
    as a LAN connection, 71
    as a WAN extension, 51
Glass house, 67
Go/no go (testing), 98
Grid (for numbering), 95
Guidelines, 154

Heuristic (versus deterministic), 63
Hired guns, 11
Hiring additional people, 163
Hop (in microwave planning), 57
Human nature, 148
Hunting license, 122

IEEE protocols, 63
Industrial quality, 88
Information interchange, 60
Install (the four step process), 5
Installation error, 153
Insurance, 97
Insurance or bonding, 110
Intent (as a legal term), 91
ISO 9000, 138
Iterative process, 159

LAN infrastructure, 152
Lasers, 54
Last minute changes, 40
Laws, Ordinances and Codes, 90
Layered approach (to design), 17
Legally sufficient, 84
Leverage, 12

Line of sight (microwave design), 57
Local laws, 11
Loop-back, 80
Losses (power), 87

Man-hours and calendar days, 162
Mandatory pre-trial settlement, 122
Marginal performers, 13
Marketing skills, 4
Mediation and arbitration, 120
Mesh diagram (as a design tool), 20
Method of payment, 158
Milestones, 168
Military (or MILSPEC quality), 88
Mission critical (requirements), 24
Mission requirement, 15
Multimode fiber optic cable, 54
Multiple networks, 62
Murphy's Law (corollary to), 136

Natural disaster, 13
Negligence or theft, 87
Negotiation:
    pretrial settlement, 122
    with vendors, 103
Negotiation skills, 4
Network address, 8
Network Layer, 62
Network management:
    applications, 4
    requirements for, 36
Non-budgeted expenses, 43
NOS (Network Operating System), 66
Numbering system, 94
Numbering systems, 8

Objective data:
    gathering, 19
    using, 103
Objective proof, 99
One unique number, 94
Open technologies, 3
Operating system support:
    client server, 67
    peer, 69
Order of precedence, 93
Ordinances, 4
OSHA (Occ. Safety & Health Act), 145
OSI (Open Systems Interconnect), 61

OSI model, 17
Other Terms and Conditions, 90
Outside consultant, 10
Outside party (for testing), 148
Outward change (organization form), 35

PAD (Packet Assembler/Disassembler), 73
Parallel Processes, 166
Parking for contractor employees, 141
Partial payments, 158
Peer Servers, 69
Peer-to-peer networks, 69
Performance bond, 121
Physical Layer, 61
PING, 79
Pirating software, 68
Plan, 5
Plenum (and cabling problems), 54
Point award structure, 107
Point values, 13
Poly Vinyl Chloride (PVC), 54
Power loss, 55
Pre-installation activities, 130
Pre-installation review, 154
Predecessor/Successor, 167
Predisposition (toward vendors), 133
Preinstallation activities, 135
Preinstallation inspection, 137
Preliminary design review, 168
Preparation (for network replacement), 9
Presentation Layer, 62
Problem solving process, 170
Project management plan, 111
Project management software, 168
Project Management Triangle, 161

Quality check, 131

Radio communications, 57
Rank ordering (respondents), 13
Real world testing, 16
Recurring expenses, 30
Redlines (on blueprints), 97
Redress grievances, 120
Redundancy, 28
References, 12
Regulations, 4

Reliance (as a legal term), 90
Remote bridge:
  extending LANs with, 72
  specification for, 92
Repeaters, 73
Reports, 177
Representative questions, 23
Residential quality, 88
Respondents' conference, 110
Retinal damage (from lasers), 79
Revisions or changes, 101
RFP (Request for Procurement), 43
RFQ (Request for Quotation), 43
Rightsizing (the organization), 2
Routers, 73
Rule of thumb:
  eliminating bidders, 108
  when checking references, 112

Safety gear, 97
Sanity check, 130
Scrap and rework, 8
Secondary path, 31
Segment:
  in prepartion of, 45
  specification for, 155
Select a consultant, 10
Self insured, 97
Self-insurers, 121
Semi-finalist (selection of), 109
Senior management and ego, 134
Server Administration, 67
Session Layer, 62
Shrinkage, 88
Simulation software (for design), 45
Single and multimode, 152
Single mode, fiber optic cable, 54
Single points of failure, 28
Sink (data), 62
Sinks and sources, 44
Site license, 24
Skills (required for effort), 5
Slippage of Schedule, 90
Sloppy workmanship, 154
Source (data), 62
Source and object code, 121
Sources to sinks, 47
Span, 113
Specification and design loop, 21
Specify (four step process), 6
Splicing (mechanical vs. fusion), 56

Staged:
    location of goods, 87
    offsite storage recommended, 138
State Statutes, 123
Statistical approach (for testing), 149
Statistically significant sample, 98
Strike, 13
Sub-contractors, 98
Sub-network, 21
Subnetting, 45
Substitute material, 155
Substituted judgment, 156
Sufficient insurance, 87
Superuser (administrator), 67
Support structure, 53
Surcharge, 144
Survey and Questionnaire, 23
Survey answers must be quantifiable,
    25

10Base2, 63
10Base5, 64
10BaseT, 64
10Broad36, 64
Technical Elegance (in design), 90
Teflon (TM), 54
Telecommunications Protocols, 67, 69
Terminating resistors, 77
Terms and conditions, 119
Test (four step process), 6
Test bed, 136
Test equipment, 76
Test for presence, 79
Test forms:
    agreement upon, 99
    proof of functionality, 150
Test plan, 76

Test the equipment, 97
Testing, 130
Testing at the application layer, 80
Testing at the physical layer, 77
The trap, 15
Theft, 145
Third Party Software (and NOSes), 67
Token, 65
TOPOGRAPHY, 49
Topologies and topographies, 2
TOPOLOGY, 49
Tradeoff (cost/bandwidth), 48
Transport Layer, 62
Tricks of the trade, 152

Uniform Commercial Code, 123
Uninsured contractors, 125
Uninterruptable power supply, 92
Unit price, 108
Unit pricing, 89
User Administration, 67, 69

Vendor tests, 81

Waivers of responsibility, 97
Walk through (of the site), 139
Weasel words, 152
What is to be delivered, 86
When it is to be delivered, 87
Who does the testing, 147
Winnowing process, 113
Work Breakdown Structure, 243

Zoning laws, 57

About the Author

Peter D. Rhodes, principal and founder of Peter Rhodes &
Associates, is a consultant specializing in LAN/WAN archi-
tectures, network management, and teleconferencing. He
holds both Bachelor's and Master's degrees in the discipline
of business management. He was pivotal in managing the
U.S. Army's telecommunications network for Headquarters,
USAREUR & 7th Army, Heidelberg, Germany. After retiring
from the Signal Corps, U.S. Army, Mr. Rhodes did systems
engineering work for The Analytic Sciences Corporation
(Reading, MA) and held a senior position with A Conference
Call, USA (Boca Raton, FL). Mr. Rhodes' previous books
include *LAN Operations: A Guide to Daily Management*. He
lives in Derry, New Hampshire.